Advances in Vibroacoustics and Aeroacustics of Aerospace and Automotive Systems

Special Issue Editors

Roberto Citarella
Luigi Federico

MDPI • Basel • Beijing • Wuhan • Barcelona • Belgrade

Special Issue Editors
Roberto Citarella
University of Salerno
Italy

Luigi Federico
Italian Aerospace Research Center (CIRA)
Italy

Editorial Office
MDPI
St. Alban-Anlage 66
Basel, Switzerland

This edition is a reprint of the Special Issue published online in the open access journal *Applied Sciences* (ISSN 2076-3417) from 2017–2018 (available at: http://www.mdpi.com/journal/applsci/special_issues/automotive_systems).

For citation purposes, cite each article independently as indicated on the article page online and as indicated below:

Lastname, F.M.; Lastname, F.M. Article title. *Journal Name* **Year**, *Article number*, page range.

First Edition 2018

ISBN 978-3-03842-851-0 (Pbk)
ISBN 978-3-03842-852-7 (PDF)

Articles in this volume are Open Access and distributed under the Creative Commons Attribution license (CC BY), which allows users to download, copy and build upon published articles even for commercial purposes, as long as the author and publisher are properly credited, which ensures maximum dissemination and a wider impact of our publications. The book taken as a whole is © 2018 MDPI, Basel, Switzerland, distributed under the terms and conditions of the Creative Commons license CC BY-NC-ND (http://creativecommons.org/licenses/by-nc-nd/4.0/).

Table of Contents

About the Special Issue Editors

Roberto Citarella, Associate Professor of Machine Design in the Department of Industrial Engineering (DIIN) at the University of Salerno. In 1996 he was Visiting Researcher at the Wessex Institute of Technology (WIT), Southampton (UK), and then appointed Fellow of WIT. In 1998 and 2000 was Visiting Researcher at the Queen Mary and Westfield College, London (UK). In 2016 was visiting professor at University of Porto, Portugal. Involved, as main investigator and member, in several national and international research activities. His main topics of research are: Boundary Element Method, Vibrational Acoustics, Bioengineering, Fracture Mechanics, Thermomechanical Fatigue. He has published nearly 140 technical papers on international peer reviewed journals and conference proceedings. He serves as editorial board member and associate editor of several international journals.

Luigi Federico, Head of the Environmental Impact of Air Transport System Unit at the Italian Aerospace Research Center (CIRA). His research activities include: Experimental and Computational Acoustics of fixed and rotating wing aircraft, Vibrational Acoustics, Boundary and Finite Element Methods, Statistical Energy Analysis, Multibody and Air Transport System Sustainability. His research unit is accredited by ENAC for noise certification according to ICAO Annex 16. Since 2018 he is contract professor of Computational Vibroacoustics at the University of Salerno. He is author of more than 50 CIRA scientific research documents and nearly 20 technical papers on international peer reviewed journals and conference proceedings. He serves as associate editor of international journals.

applied sciences

MDPI

Editorial

Advances in Vibroacoustics and Aeroacustics of Aerospace and Automotive Systems

Roberto Citarella [1,*] **and Luigi Federico** [2]

[1] Department of Industrial Engineering, University of Salerno, 84084 Fisciano, Italy
[2] Italian Aerospace Research Centre (C.I.R.A), via Maiorise snc, 81043 Capua, Italy; l.federico@cira.it
* Correspondence: rcitarella@unisa.it

Received: 23 February 2018; Accepted: 27 February 2018; Published: 3 March 2018

1. Introduction

Computing sound, taking into account the dynamical phenomena leading to its generation and transmission through fluid, is a challenging issue, with several applications into different industrial sectors [1,2]. Automotive and aerospace industries, dealing with vehicles and their comfortability for passengers' transportation and payload preservation, constitute two natural frameworks in which these subjects acquire a prominent role.

In principle, aeroacoustics calculations can be addressed solving Navier–Stokes equations, exploiting direct numerical simulations. In practice, such approach is extremely time-consuming, leading to the introduction of the analogy-based approaches, traditionally ascribed to the field of computational aeroacoustics (CAA). With this kind of method, the near field aerodynamics is computed, giving velocity and pressure fluctuations used as acoustic source terms [3].

Aeroacoustic noise represents a boundary condition applied on the vehicle model, in applications of industrial interest. Recently, the interest in such a kind of calculations is grown, thanks to the enhanced computational power of nowadays computers. In [4], for instance, a turbulent model was used in the CFD calculations and the rear of a car model was optimized using genetic algorithm method. This is an example of the increasing interest in detailed aeroacoustics simulations at the design level in automobiles.

Alongside with aeroacoustics, vibration studies cover a crucial role in applications [5,6]. Several deterministic and statistical approaches exist, and each of them bears advantages and limitations. The challenge when using deterministic methods in case of complex application systems derive from the need for accurate vibrational response of the model. Validating each mode of a complex structure, when the frequency range of interest covers hundreds of natural frequencies is computationally expensive. In addition, in order to have an accurate response, the model cannot be oversimplified and this adds a significant cost in terms of computation. The advantage of using statistical methods derives from simpler requirements for validating each eigenfrequency, even if these approaches are less accurate for a specific solution with a clear tonal resonance.

For these reasons, hybrid methods are appealing, merging together the strength point of both deterministic and statistical simulation techniques. Lower frequencies—where the tonal resonances are significant—are calculated applying finite element methods, whereas for higher frequencies, a statistic energy approach can be chosen.

The idea of identifying noise paths in vehicle noise transmission using statistical energy analysis (SEA) is not new, being already established in 1980s by DeJong. In [7], DeJong created a SEA model of a passenger car and used it to identify the transmission paths of acoustic energy from the engine, tires, and floor to the interior of the vehicle, predicting the noise level of the passenger compartment. In recent years, this idea has been extended: Musser and coworker [8], for instance, simulated vehicle interior noise levels using SEA, validating the study with a full sized car.

On the other hand, the finite element method has gained attention, as in [9] where a finite element structural acoustic model of a car has been made.

As an alternative to FEM, the boundary element method (BEM) can be adopted for noise emission calculation [10,11]. In [12], the numerical modelling of noise radiated by a car body, using the so-called acoustic transfer vector (ATVs) and modal acoustic transfer vector (MATVs) concepts, is presented.

Combination of the aeroacoustics and vibroacoustics simulation analysis lies at the core of the current applied research in transportation industries. Müller and coworkers [13] used fast multipole BEM (FMBEM) to characterize the exterior aeroacoustic loads of a vehicle, combined then to statistical energy analysis simulation. Cotoni and coworkers [14] used FEM combined to statistical energy analysis simulation.

Concerning the FMBEM for the exterior problem, there still remains the famous issue of the so-called irregular or spurious frequencies, especially in the mid-frequency domain in which there is a high modal density.

Wang and coworkers [15] calculated a car interior noise level by combining the LES (large eddy simulation) to FEM/BEM simulation, validating CFD results in a wind tunnel by 1/3rd size model.

In summary, numerous newer strategies have been proposed in recent years, circumventing some of the problems and sources of inefficiency found in the past. The aim of this special issue is to help to shed some light on current progress in applied research in the field of transportation, linking the results obtained in recent years with industrial development, in the pursuit towards attaining more and more comfortable and sustainable mobility.

2. The Present Issue

We are deeply honored to be the guest editors of this thematic issue, which includes original articles and reviews of interest for the scientific community working on these subjects, as well as for industrial R&D departments. We are particularly grateful to the several leaders in these areas who decided to honor us with their valuable contributions. We want to thank them and their collaborators for the effort they have done for publishing their works in this issue of *Applied Sciences*. We are equally thankful to many colleagues who have helped us evaluating and reviewing these manuscripts with substantial improvements. In particular, we appreciate the editor-in-chief of *Applied Sciences* for inviting us to edit this topic issue and the staff of the editorial office who supported the entire process.

The purpose of this special issue is to highlight the latest advances in vibroacoustics and aeroacoustics applied to aerospace and automotive industries. In proposing this wide topic, we wanted to give scientists and engineers the opportunity to publish their studies of current interest, both in the theoretical and experimental fields of vibration, sound generation, and radiation; and passive and active noise control approaches, which can be of interest for assessing vibroacoustics characteristics of aerospace and automotive vehicles. In particular, we have tried to include articles focusing on the introduction of new approaches and methodologies in the vibroacoustics and aeroacoustics fields.

This issue provides access to several articles encompassing the aforementioned themes.

In the first article [16], an innovative integrated design verification process is described, based on the bridging between a new semiempirical jet noise model and a hybrid finite-element method/statistical energy analysis (FEM/SEA) approach, for calculating the acceleration produced at the payload and equipment level within the structure of Vega-C Launcher, vibrating under the external acoustic forcing field. The result is a verification method allowing for accurate prediction of the vibroacoustics in the launcher interior, using limited computational resources and without resorting to computational fluid dynamics (CFD) data.

Another paper of this issue [17], estimated the internal aerodynamic noise due to valve flow in a simple constriction-expansion pipe (a pressure relief device), by combining the large eddy simulation technique with a wavenumber-frequency analysis, which made it possible to decompose the fluctuating pressure into the incompressible hydrodynamic pressure and compressible acoustic pressure. The results showed that the acoustic pressure fluctuations in a pipe could be separated from the incompressible ones. This made it possible to obtain accurate information about the acoustic power, which could be used to assess the likelihood of a piping system failure due to acoustic-induced vibration, along with information about the acoustic power spectrum of each acoustic mode, which could be used to facilitate the systematic mitigation of the potential acoustic-induced vibration in piping systems.

In [18], flow-induced vibrations and the sound radiation of flexible plate structures of different thickness mounted in a rigid plate were experimentally investigated. Therefore, flow properties and turbulent boundary layer parameters were determined through measurements with a hot-wire anemometer in an aeroacoustic wind tunnel. Furthermore, the excitation of the vibrating plate was examined by laser scanning vibrometry.

In [19], fully-resolved rotor-fuselage interactional aerodynamics was used as the forcing term in a model based on the Euler–Bernoulli equation, aiming to simulate helicopter tail-boom vibration. The model was based on linear beam analysis and captured the effect of the blade-passing as well as the effect of the changing force direction on the boom. Results for the tail-boom vibration served to demonstrate the strong effect of aerodynamics on tail-boom aeroelastic behavior.

In [20], the mechanism by which plasma actuators can control the cavity tone was explored; in particular, the amount of sound reduction ('control effect') produced by actuators of differing dimensions was measured and direct aeroacoustic simulations were performed. The simulations showed that longitudinal streamwise vortices are introduced in the incoming boundary by the actuators, and the vortices form rib structures in the cavity flow. These vortices distort and weaken the two-dimensional vortices responsible for producing the cavity tone, causing the tonal sound to be reduced.

In [21], the prediction and reduction of noise from subsonic jets through the reconstruction of turbulent fields from Reynolds-averaged Navier–Stokes (RANS) calculations are addressed. This approach, known as stochastic noise generation and radiation (SNGR), reconstructs the turbulent velocity fluctuations by RANS fields and calculates the source terms of vortex sound acoustic analogy. An active fluid injection technique, based on extractions from turbine and injections of high-pressure gas into the main stream of exhausts, has been proposed and finally assessed with the aim of reducing the jet-noise through the mixing and breaking of the turbulent eddies. The SNGR method is, therefore, suitable to be used for the early design phase of jet-noise reduction technologies.

In [22], an aerodynamic noise optimization was performed by a non-linear acoustics solver (NLAS) approach for acoustic calculation. With the use of Kriging surrogate model, a multi-objective optimization of the streamlined shape of high-speed trains was obtained, which takes the noise level in the far field and the drag of the whole train as the objectives. To efficiently construct the Kriging model, the cross validation approach has been adopted.

In [23], a vibro-acoustic numerical and experimental analysis was carried out for the chain cover of a low powered four-cylinder four-stroke diesel engine. By applying a methodology used in the acoustic optimization of new engine components, firstly a finite element model (FEM) of the engine was defined, then a vibration analysis was performed for the whole engine (modal analysis), and finally a forced response analysis was developed for the only chain cover (separated from the overall engine). A boundary element (BE) model of the only chain cover was realized to determine the chain cover noise emission, starting from the previously calculated structural vibrations. All the information thus obtained allowed the identification of those critical areas, in terms of noise generation, in which to undertake necessary improvements.

The paper [24] provides a summary of some common research works carried out by the authors concerning computational methods for the prediction of the responses in the frequency domain of general linear dissipative vibroacoustics (structural-acoustic) systems for liquid and gas in the low-frequency (LF) and medium-frequency (MF) domains, including uncertainty quantification (UQ) that plays an important role in the MF domain. An efficient reduced-order computational model (ROM) is constructed using a finite element discretization (FEM) for the structure and the internal acoustic fluid. The external acoustic fluid is treated using a symmetric boundary element method (BEM) in the frequency domain. All the required modeling aspects required for the analysis in the MF domain have been introduced, in particular the frequency-dependent damping phenomena and model uncertainties.

Conflicts of Interest: The authors declare no conflict of interest.

References

1. Fahy, J.; Gardonio, P. *Sound and Structural Vibration*, 2nd ed.; Academic Press: Cambridge, MA, USA, 2007.
2. Ohayon, R.; Soize, C. *Advanced Computational Vibroacoustics*; Reduced Order Models and Uncertainty Quantification; Cambridge University Press: Cambridge, UK, 2014.
3. Wagner, C.; Huttl, T.; Sagaut, P. *Large-Eddy Simulation for Acoustics*; Cambridge University Press: Cambridge, UK, 2007.
4. Beigmoradi, S.; Hajabdollahi, H.; Ramezani, A. Multi-objective aeroacoustic optimization of rear-end in a simplified car model by using hybrid Robust Parameter Design, Artificial Neural Networks and Genetic Algorithm methods. *Comput. Fluids* **2013**, *90*, 123–132. [CrossRef]
5. Siano, D.; Citarella, R. Elastic Multi Body Simulation of a Multi-Cylinder Engine. *Open Mech. Eng. J.* **2014**, *8*, 157–169. [CrossRef]
6. Armentani, E.; Sbarbati, F.; Perrella, M.; Citarella, R. Dynamic analysis of a car engine valve train system. *Int. J. Veh. Noise Vib.* **2016**, *12*, 229–240. [CrossRef]
7. DeJong, R. *A Study of Vehicle Interior Noise Using Statistical Energy Analysis*; SAE Technical Paper 850960; SAE: Warrendale, PA, USA, 15 May 1985.
8. Musser, C.; Manning, J.; Peng, G. Predicting Vehicle Interior Sound with Statistical Energy Analysis. *Sound Vib.* **2012**, *46*, 8–14.
9. Durand, J.F.; Soize, C.; Gagliardini, L. Structural-acoustic modeling of automotive vehicles in presence of uncertainties and experimental identification and validation. *J. Acoust. Soc. Am.* **2008**, *124*, 1513. [CrossRef] [PubMed]
10. Citarella, R.; Landi, M. Acoustic analysis of an exhaust manifold by Indirect Boundary Element Method. *Open Mech. Eng. J.* **2011**, *5*, 138–151. [CrossRef]
11. Armentani, E.; Trapani, R.; Citarella, R.; Parente, A.; Pirelli, M. FEM-BEM Numerical Procedure for Insertion Loss Assessment of an Engine Beauty Cover. *Open Mech. Eng. J.* **2013**, *7*, 27–34. [CrossRef]
12. Citarella, R.; Federico, L.; Cicatiello, A. Modal acoustic transfer vector approach in a FEM–BEM vibro-acoustic analysis. *Eng. Anal. Bound. Elem.* **2007**, *31*, 248–258. [CrossRef]
13. Müller, S.; Cotoni, V.; Connelly, T. Guidelines for Using Fast Multipole BEM to Calculate Automotive Exterior Acoustic Loads in SEA Models. *SAE Int. J. Passeng. Cars Mech. Syst.* **2009**, *2*, 1530–1537. [CrossRef]
14. Cotoni, V.; Shorter, P.J.; Langley, R.S. Numerical and experimental validation of a hybrid finite element-statistical energy analysis method. *J. Acoust. Soc. Am.* **2007**, *122*, 259–270. [CrossRef] [PubMed]
15. Wang, Y.; Zhen, X.; Wu, J.; Gu, Z.; Xiao, Z.; Yang, X. Hybrid CFD/FEM-BEM simulation of cabin aerodynamic noise for vehicles traveling at high speed. *Sci. China Technol. Sci.* **2013**, *56*, 1697–1708. [CrossRef]
16. Bianco, D.; Adamo, F.P.; Barbarino, M.; Vitiello, P.; Bartoccini, D.; Federico, L.; Citarella, R. Integrated Aero–Vibroacoustics: The Design Verification Process of Vega-C Launcher. *Appl. Sci.* **2018**, *8*, 88. [CrossRef]
17. Kim, K.; Ku, G.; Lee, S.; Park, S.; Cheong, C. Wavenumber-Frequency Analysis of Internal Aerodynamic Noise in Constriction-Expansion Pipe. *Appl. Sci.* **2017**, *7*, 1137. [CrossRef]
18. Osterziel, J.; Zenger, F.J.; Becker, S. Sound Radiation of Aerodynamically Excited Flat Plates into Cavities. *Appl. Sci.* **2017**, *7*, 1062. [CrossRef]
19. Batrakov, A.; Kusyumov, A.; Kusyumov, S.; Mikhailov, S.; Barakos, G.N. Simulation of Tail Boom Vibrations Using Main Rotor-Fuselage Computational Fluid Dynamics (CFD). *Appl. Sci.* **2017**, *7*, 918. [CrossRef]
20. Yokoyama, H.; Tanimoto, I.; Iida, A. Experimental Tests and Aeroacoustic Simulations of the Control of Cavity Tone by Plasma Actuators. *Appl. Sci.* **2017**, *7*, 790. [CrossRef]
21. Barbarino, M.; Ilsami, M.; Tuccillo, R.; Federico, L. Combined CFD-Stochastic Analysis of an Active Fluidic Injection System for Jet Noise Reduction. *Appl. Sci.* **2017**, *7*, 623. [CrossRef]

22. Sun, Z.; Zhang, Y.; Yang, G. Surrogate Based Optimization of Aerodynamic Noise for Streamlined Shape of High Speed Trains. *Appl. Sci.* **2017**, *7*, 196. [CrossRef]
23. Armentani, E.; Sepe, R.; Parente, A.; Pirelli, M. Vibro-Acoustic Numerical Analysis for the Chain Cover of a Car Engine. *Appl. Sci.* **2017**, *7*, 610. [CrossRef]
24. Ohayon, R.; Soize, C. Computational Vibroacoustics in Low- and Medium- Frequency Bands: Damping, ROM, and UQ Modeling. *Appl. Sci.* **2017**, *7*, 586. [CrossRef]

© 2018 by the authors. Licensee MDPI, Basel, Switzerland. This article is an open access article distributed under the terms and conditions of the Creative Commons Attribution (CC BY) license (http://creativecommons.org/licenses/by/4.0/).

applied sciences

MDPI

Article

Integrated Aero–Vibroacoustics: The Design Verification Process of Vega-C Launcher

Davide Bianco [1,*], Francesco P. Adamo [1], Mattia Barbarino [1], Pasquale Vitiello [1], Daniele Bartoccini [2], Luigi Federico [1] and Roberto Citarella [3]

1 Italian Aerospace Research Centre (C.I.R.A), via Maiorise snc, 81043 Capua, Italy; f.adamo@cira.it (F.P.A.); m.barbarino@cira.it (M.B.); p.vitiello@cira.it (P.V.); l.federico@cira.it (L.F.)
2 Avio S.p.A., I-00034 Colleferro, Italy; d.bartoccini@avio.it
3 Department of Industrial Engineering, University of Salerno, 84084 Fisciano (SA), Italy; r.citarella@unisa.it
* Correspondence: d.bianco@cira.it; Tel.: +39-0823-62-3572

Received: 21 November 2017; Accepted: 5 January 2018; Published: 10 January 2018

Abstract: The verification of a space launcher at the design level is a complex issue because of (i) the lack of a detailed modeling capability of the acoustic pressure produced by the rocket; and (ii) the difficulties in applying deterministic methods to the large-scale metallic structures. In this paper, an innovative integrated design verification process is described, based on the bridging between a new semiempirical jet noise model and a hybrid finite-element method/statistical energy analysis (FEM/SEA) approach for calculating the acceleration produced at the payload and equipment level within the structure, vibrating under the external acoustic forcing field. The result is a verification method allowing for accurate prediction of the vibroacoustics in the launcher interior, using limited computational resources and without resorting to computational fluid dynamics (CFD) data. Some examples concerning the Vega-C launcher design are shown.

Keywords: transport; noise; acoustics; rocket; analysis

1. Introduction

Many factors influence the definition and selection of the structural design concept. In general, in complex systems, structural designing is an iterative process. The process starts with the conceptual design of possible alternatives, which could be considered to satisfy the general performance requirements and are likely to meet the main mission constraints (e.g., mass, interfaces, operation and cost).

The system's engineering activities are equally valid and necessary at all levels of decomposition within the space product. In particular, the verification engineering function, which iteratively compares the outputs from other functions with each other, in order to converge upon satisfactory requirements, functional architecture, and physical configuration, defines and implements the processes by which the finalized product design is proved to be compliant with its requirements.

According to the European Space Agency (ESA) standards [1], the verification process activities shall be incrementally performed at different levels and in different stages, applying a coherent bottom-up building-block concept and utilizing a suitable combination of different verification methods. To reach the verification objectives, a verification approach shall be defined in an early phase of the project by analyzing the requirements to be verified, taking into account the following:

- design peculiarities;
- the qualification status of the candidate solution;
- the availability and maturity of verification tools;
- verification and test methodologies;
- programmatic constraints;

- the cost and schedule.

In the mechanical design of launchers, the verification of vibroacoustics requirements is crucial for guaranteeing the preservation of the payloads' functionalities during the preorbital phase. This process is particularly affected by the lack of both the availability and maturity of standardized design testing tools. This work describes the innovative integrated process used at the design-stage level in the verification of the Vega-C rocket project development.

The interstages of the new *Vettore Europeo di Generazione Avanzata* (VEGA, European Advanced Generation Vector) configurations are examined in the framework of the ESA-funded VEGA Consolidation and Evolution Program (VECEP), in order to derive the acceleration level at the equipment locations and the average vibration response of the connecting interfaces. The analysis is needed in the framework of the re-engineering phase aimed at further increasing the performance of the VEGA launcher, and it is part of a preliminary numerical verification necessary to assess the compliance of the designed structure with the acceleration level imposed as a limit for the protection of the equipment and payload.

The complexity of the analysis is driven by two critical elements, which have been addressed in this work with an innovative approach. First, the acoustic pressure generated at lift-off cannot be calculated with the resolution needed because of the computational complexity of the simulation. Then, the vibroacoustic analysis driven by the calculated acoustic field on the basis of finite-element/boundary-element method (FEM/BEM) techniques turns out to be unfeasible in a medium to high frequency range, even when resorting to computationally efficient approaches such as the acoustic transfer vector (ATV) or modal ATV (MATV) [2], because of the high modal density of the structure. On the other hand, the employment of a fully energy-based approach in this frequency range is restricted by the need to simulate the equipment as lumped masses, connected to the main structure throughout rigid links.

In this paper, we propose an integrated verification approach [3–5] exploiting an innovative semiempirical Eldred-based source model with BEM propagation [6,7] for building the jet acoustic pressure field, as well as a state-of-the-art hybrid FEM/statistical energy analysis (SEA) method adopted to combine the equipment's local deterministic responses with the mean value of the dynamic response of the launcher's major sections. In Figure 1, the logical structure of the whole integrated method is shown, with the acoustic field calculation feeding the structural dynamic simulation.

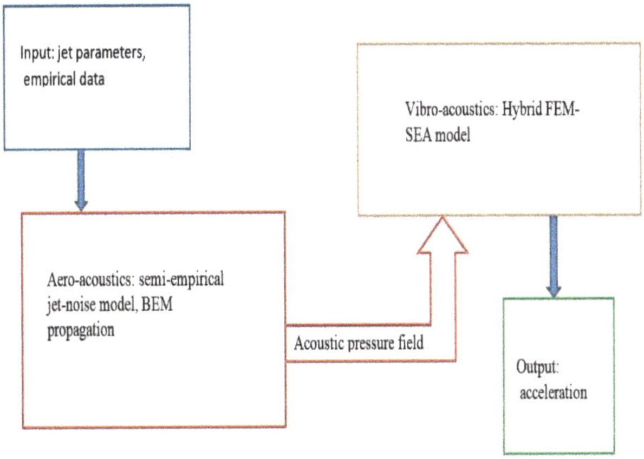

Figure 1. The logical scheme of the integrated multidisciplinary approach used for the VEGA Consolidation and Evolution Program (VECEP) design verification process, described in this paper, is shown.

The power of the proposed approach, with respect to other hybrid vibroacoustic methods, lies in its integration with the aeroacoustic component. In fact, a realistic model for the forcing pressure clearly implies a realistic reconstruction of the structural response in the operating condition considered.

The concept of a space launcher can be, in some sense, elusive. It is thus worthy to try to define the range of applicability of the approach described in this work. In particular, the model used for reconstructing the acoustic pressure field acting on the structure and due to jet fluid dynamics aspects is general and independent from the type of the structure taken into account as well as from considering a horizontal or vertical launch. On the other hand, the hybrid techniques describing the response of the structure to the load cannot be reliable for geometries including wings, because of the complications arising in the model and the possible fall down of the diffuse-field assumption.

The semiempirical approach for reconstructing the acoustical pressure component of the rocket jet is described in Section 2. It is an improvement of the Eldred Standard model, based on heuristic assumptions and scaling laws, properly tuned against experimental data from different rocket engines [8–11]. In this simple model, in order to determine a realistic equivalent noise source, several parameters, such as the vehicle geometry or launch pad configurations, have to be taken into account. Several hybrid methods have been proposed for overcoming this limit [12,13], and the new formulation contains an explicit expression for the acoustic pressure of each noise source, in terms of amplitude and phase, in order to account for correlation effects and propagate the signal using a wave equation. The noise prediction obtained with the revised Eldred-based model has thus been used for formulating an empirical/BEM approach that allows us to evaluate the scattering effects, which, summed with the direct, correlated acoustical pressure, is exploited for the prediction of the aeroacoustics loads of the VEGA launch vehicle at lift-off.

The hybrid FEM/SEA vibroacoustic approach is described and validated in Section 3. The implemented analysis resorts to a FEM solver to extract the modal parameters associated to the deterministic subsystems of the VEGA interstages. The development of the statistical subsystems and the set-up of the global hybrid models are realised within a numerical environment, formerly devoted to SEA methodology only, in which the required dynamic analysis is also performed. The structural and acoustical models of the interstages are validated through a comparison with theoretical and numerical estimates. As required for the assessment of the equipment qualification status, the activity is finalized with the delivery of the dynamic response database, numerically generated by applying the acoustic pressure field to the interstage validated models.

Two sections, dealing with the discussion of the results and some concluding remarks, close the paper.

2. Jet-Noise Source Model and Propagation

The complexity of the detailed modeling of the acoustic component in rocket exhaust can be circumvented using empirical approaches, on the basis of the assumption that an equivalent source distribution can be determined using few parameters (e.g., jet exit conditions and acoustic efficiency) and some universal curves determined experimentally. In particular, the first Eldred model [8] is based on the observation that each slice of the jet contributes to acoustic radiation, mostly in the frequency band f_i that depends on the distance of the slice from the exhaust. The subsequent reformulation of the approach states that each slice contributes to the full noise spectral frequency content with an overall power that depends on its curvilinear distance from the exhaust [12]. Differently from the original Eldred model, which directly provides a definition of the sound pressure level (SPL), its reformulation can be revised to provide an explicit formulation of the acoustic pressure phasor p_j [14]. In particular, this simpler version of the reformulated semiempirical model moves from the relation

$$p_{j0} = g(r_j)\sqrt{P_0 \cdot 8\pi\rho c} \tag{1}$$

equating the phasor of the incident pressure, p_{j0}, due to the jth source, to the corresponding monopole amplitude, $Q_0 = \sqrt{P_0 \cdot 8\pi\rho c}$, with P_0 indicating the power, ρ being the air density and c being the speed of sound at ambient conditions, through the green function of the nonconvective Helmholtz operator, $g(r_j)$; the directivity effect produced by the inflow refraction, can be taken into account with a directivity function $D = \frac{I}{I_0}$, where $I_0 = \frac{P_0}{4\pi r^2}$ and I indicates the intensity. Hence, the pressure of each monopole source, composing the equivalent set that reproduces the jet noise, can be expressed in the form

$$p_j = g_j(r)Q_0\sqrt{D_j(\theta)} \tag{2}$$

with θ indicating the angle between the observer and the monopole source position.

The second formulation of the Eldred model assumes a contribution from each jet slice to the full noise spectral frequency content with an overall power that depends on its curvilinear distance from the exhaust [12]. Denoting $w(s_k)$ as the sound power per unit axial length, the contribution of each slide reads as

$$w(s_k)\Delta s_k = \frac{L_{ref}w(s_k)}{W_{ac}}W_{ac}\frac{\Delta s_k}{L_{ref}} \tag{3}$$

where L_{ref} is a jet reference length, and k denotes the kth slice at a distance s_k from the exhaust. W_{ac} represents the overall acoustic power [15] of the engine, which is a fraction of the mechanical power W_m ($W_{ac} \simeq \eta W_m$ with $\eta = 0.03$).

The overall power density level, $w(s_k)$, can be distributed on the frequency spectrum. Denoting with $w(s_k, f_i)$ the sound power level per unit axial length and unit frequency, one can introduce the reduced sound power density:

$$\tilde{\tilde{w}} = \frac{w(s_k, f_i)}{w(s_k)}\frac{u_j}{s_k}\frac{c_a}{c_j} \tag{4}$$

with u_j being the jet speed at the exhaust, c_a being the ambient sound speed and c_j being the jet exhaust speed. Consequently, it is possible to obtain the monopole source's acoustic power at each axial length and for each frequency from

$$P_{ik} = \tilde{\tilde{w}} \cdot \tilde{w} \cdot W_{ac} \cdot \frac{s_k}{u_j}\frac{c_j}{c_a}\frac{\Delta s_k \Delta f_i}{L_{ref}} \tag{5}$$

The revised Eldred-based model is enriched with two free parameters, α and β, respectively related to the mixing-layer turbulence length scale and its integral correlation length.

The Eldred standard model describes the jet noise with a set of equivalent sources, laying along the jet axis with no consideration of their number or spatial distribution. It is reasonable to think that the spatial scale of each source, and consequently its number, is related to its turbulent length scale. In analogy with similar physical consideration taken from the Stochastic Generation and Radiation model for turbulent Noise (SNGR) [16], the energy peak of the universal energy spectrum can be related to the turbulent length scale through a free mode parameter, according to

$$L_T = \frac{s}{1.5} \cdot \frac{1}{\alpha} \tag{6}$$

This equation puts in relation the turbulent scale L_T with the value of the curvilinear abscissa s, and takes into account the experimental evidence according to which the energy peak has its maximum at a value 1.5 of the modified Strouhal number [14]. Equation (6) states that the turbulent scale grows up gradually increasing the distance from the jet nozzle, with sources emitting at lower frequencies at a greater distance from it.

On the other hand, a classical result [17] relates the integral length scale L_{int} for homogeneous, isotropic turbulence with the correlation length $L_0 = \frac{2\pi}{k_0}$ of a vortex structure at a single wavenumber k_0, according to the following equation:

$$L_{int} = \frac{3}{8}L_0 \tag{7}$$

This equation implies then that the correlation length at each frequency is proportional to the wavelength. In the revised Eldred-based model, it would means that each single source is correlated with all the others with a correlation length that depends on the frequency considered. In order to enhance the capabilities of the revised model and in order to tune the model exploiting the experimental data, the parameter β is then added to the equation:

$$L_{int} = \frac{3}{8} L_0 \cdot \frac{1}{\beta} \quad (8)$$

A comprehensive analysis concerning the role of β is provided in [14].

Both Eldred-based uncorrelated and correlated source models can be used for the hybrid empirical/BEM computation. The hybrid approach's implementation has firstly been tested to check that the total pressure, p, is equal to the Eldred incident field, p_{inc}, when a vanishing scattering BEM surface is considered ($p_{sc} = 0$).

3. Hybrid FEM/SEA Model and Interstage Response Analysis

As pointed out in the introduction, the objective of this work is to predict the acceleration response level at the equipment and payload location, for subsequent verification of their qualification status. In addition, the dynamic average response of the VEGA interstage interfaces IS01, IS12 and IS23 to the acoustic field generated at lift-off by the P120 engine has been calculated with the innovative approach described in the previous section.

A numerical prediction has been obtained through the development of hybrid FEM/SEA models and their analysis in the 100–1600 Hz frequency range using the commercial code VAOne [18], in order to overcome the limitation of deterministic tools in the investigated frequency range. The fourth section of VEGA, constituted by the interstage IS34, the Attitude Vernier Upper Module (AVUM) and the Fairing, was left out from the numerical investigation because it has been unchanged with respect to the old configuration. The large frequency range spread of the pressure power spectra motivates the need for a vibroacoustic response analysis accounting for all the frequency content of the main external pressure field to avoid underestimating the equipment's and components' acceleration levels. Despite the relatively high skin thickness and the opportunity to perform the analysis on each component separately, the large dimensions of the VEGA interstages still require the involvement of a large number of structural modes, not to mention at all the even higher modal density of the coupled interior acoustic cavities.

As a consequence, above 400/500 Hz, the FEM technique turns out to be computationally unfeasible even when resorting to modal superposition procedures, and the SEA approach becomes more appropriate for acquiring the responses of both the structure and the acoustic cavity. As a matter of fact, SEA represents systems in terms of a number of coupled subsystems, taking into account the global energy trade-off due to their interaction, namely, energy storage, transmission and dissipation. The parameters in SEA equations are typically obtained with certain statistical assumptions about the dynamic properties of each subsystem. For running the calculations, the commercial code VAOne has been used.

The code allows for a complete range of analysis because different solution methods and modules are available, specifically, FEM, BEM and SEA. The capabilities in dealing with any kind of dynamic problem are even enlarged by the option to develop hybrid models, in which all available methodologies are combined. The development of structural models entirely based on a SEA approach was restricted by the need to simulate the equipment as a lumped mass, connected to the main structure throughout rigid links. These components are not suitable for SEA modeling because they have no modal behavior by definition, while a large modal participation is required for a SEA model to be accurate.

The adoption of a hybrid approach was mandatory for the monitoring of the equipment acceleration levels, in order to verify their qualification status. Consequently, for all interstages, a full SEA model could not be developed, and a careful identification of the number and extension of the interstage parts to be modeled as FEM subsystems was required to avoid building up a model unsuitable for a SEA approach. Moreover, all the external and internal interfaces have been modeled as FEM subsystems to retain all their dynamic contributions, otherwise ignored by using a SEA "beam"-type subsystem for their simulation.

The correct balance among the different type of subsystems (either SEA or FEM) was essential to allow a fairly quick analysis with no penalties in terms of accuracy. General rules to check the accuracy of a SEA subsystem are based on the following concepts, applicable to the frequency bands chosen for the analysis:

- A mode count of no less than 5.
- A model overlap factor (MOF) greater than or equal to the unitary value.

Both parameters depend on the modal density, representing the number of modes per hertz associated to the subsystem, while the MOF is also dependent on the loss factor; together they provide the conditions for which a "diffuse-kind" vibration or acoustic field characterizes the SEA subsystem's dynamic behavior.

Hybrid FEM/SEA modeling, then, requires an iterative process aiming for the exact identification of the frequency range accuracy for each subsystem in order to correctly define their geometries and dimensions. For accurate predictions in a large frequency band, more than one model might be necessary, where the number of subsystems modeled according to one methodology or the other could consequently change. The validation process of the SEA and hybrid structural models is carried out by relying on the following:

- Theoretical assessments of SEA subsystems against simplified formulations to verify the consistency with some global distinguishing parameters.
- Numerical checks on sections of the interstages performed by comparison of the prediction data provided by the hybrid model and by the equivalent FEM model.
- Numerical mutual checks between the full interstage hybrid and FEM models performed in the overlap frequency region where the two approaches are reasonably applicable.

The acoustic loading applied is considered a diffuse field, acting on the external surface of interstages; the power spectral density (PSD) of the pressure autocorrelation function is assumed to be constant over the entire surface.

The responses of all the interstages are expected to be characterized by a more or less uniform modal energy, which is beneficial to reduce the effect of nonideal modeling, in terms of SEA accuracy, of some of the model components, and in order to reduce the effect of not simulating the entire launch vehicle.

In fact, modal energy equivalence can also be interpreted as an opportunity to perform a first approximation analysis on each component or subsystem by assuming it as being disjointed from the others. This assumption is strictly applicable to models made of SEA subsystems only and for which SEA criteria are fulfilled; thus, the application to hybrid models must be carefully verified, because of the "non-diffuse-field" characteristics of FEM subsystems.

The Interstage Models

The interstages IS01 (Figure 2) and IS12 are made of aluminium. The former is cylindrical and presents a nonuniform lateral surface thickness, 4 mm thicker in the area in which the openings are located, whereas the latter is constituted by two sections, having the form of a cone frustum connected through an internal interface. The third interstage, IS23, has different characteristics; it is realized by a composite material, and it is reinforced over the lateral surface by axial and circumferential stiffeners. The shape is still conical, split into two sections connected by an internal frame. Frames, reinforcing elements and skins are all constituted of a composite material.

The simpler VEGA section among those analyzed, namely, interstage IS01, is taken as an example for illustrating the final part of the design verification described in this paper. Its main geometrical and dynamic characteristics are summarized in Table 1, along with the theoretical estimates of the ring and coincidence frequencies referred to the mean thickness.

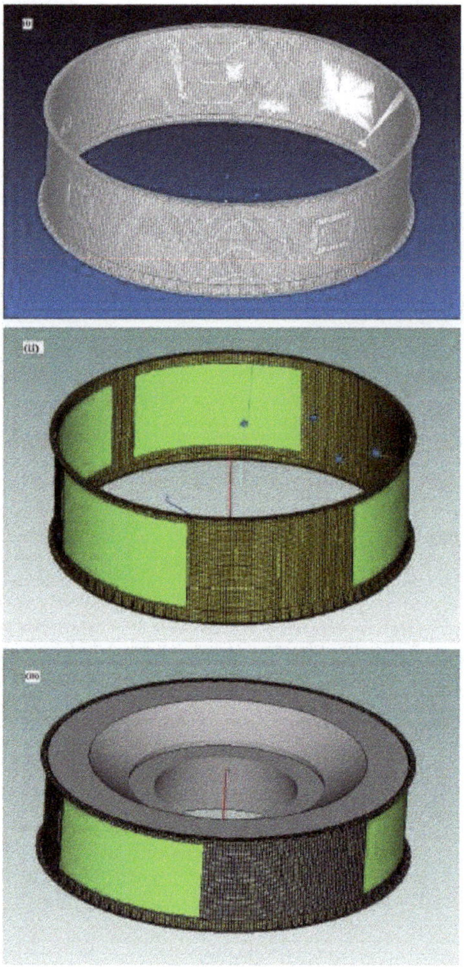

Figure 2. From the top to the bottom: (**i**) finite-element method (FEM) model; (**ii**) hybrid model; and (**iii**) acoustic coupled model for the IS01 interstage.

Table 1. Geometrical characteristics and theoretical estimates of dynamic frequencies for the IS01 interstage.

Property	**Value**
Length (m)	1.18
Diameter (m)	3.4
Ring freq. (Hz)	500
Coincidence freq. (Hz)	1100

The various subsystems of the interstage model are shown in Figure 2: the SEA structural subsystem in green, the FEM structural subsystem in gold–brown, and the SEA acoustic subsystem in grey. In the same figure, the original FEM model used for deriving the hybrid model has been reported.

In particular, the hybrid FEM/SEA model has been built up by assembling six FEM structural subsystems (two large subsystems for the upper and lower rings, two large subsystems for the areas in which several equipment are located, and two small subsystems, each connected to a single equipment component), four large SEA structural subsystems (modeling the areas with lower thickness) and one SEA acoustic subsystem.

A preliminary check of the model was performed by analyzing the modal density and radiation efficiency of the interstage: all SEA subsystems in the IS01 hybrid model presented a good agreement with the characteristic theoretical parameters reported in Table 1 and their asymptotic values, as shown in Figure 3, where the modal density and radiation efficiency are reported for each SEA subsystem of the hybrid model.

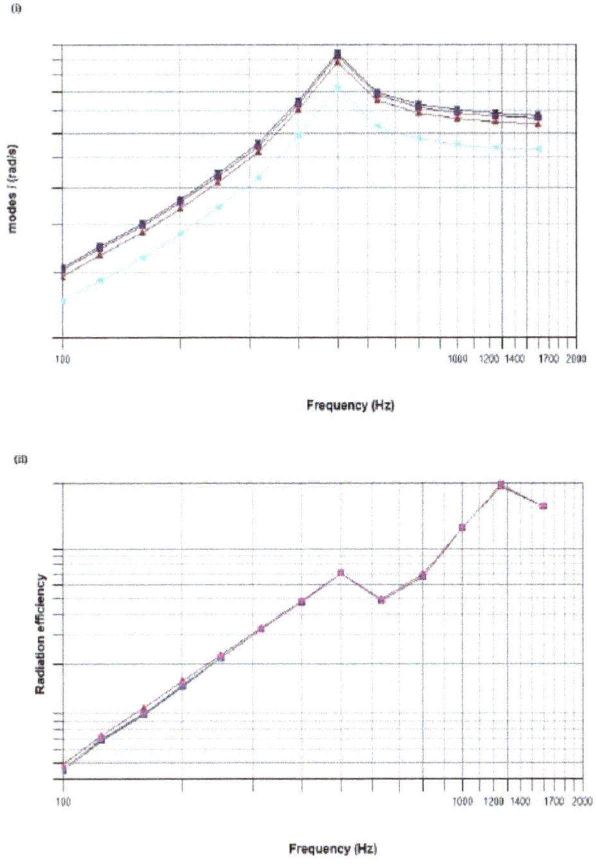

Figure 3. From the top to the bottom: (**i**) modal densities; and (**ii**) radiation efficiency for the statistical energy analysis (SEA) subsystems in the IS01 interstage hybrid model.

For each SEA subsystem, the modal density displays how the ring frequency is close to the theoretical value. Asymptotic values also confirm theoretical modal density estimates, evaluated for the curved subsystems through the equivalent flat panel relation, valid above the ring frequency f_r. This frequency, along with the structural modal density $n_s(f)$, can be expressed as

$$f_r = \frac{1}{2\pi r}\sqrt{\frac{E}{\rho - (1-\nu^2)}} \quad (9)$$

$$n_s(f) = \frac{A}{2}\sqrt{\frac{12\rho_s(1-\nu^2)}{Eh^2}}; A = \Delta\theta r L \quad (10)$$

in which the structural properties (elasticity modulus E, Poisson ratio ν, volume density ρ_s, and thickness h) are recalled along with the inner ring radius r, the axial length L and the arc angle $\Delta\theta$ [19]. The radiation efficiency curves confirm the theoretical values for both the characteristic frequencies (ring and coincidence), as they superimpose almost perfectly with discrepancies limited to low frequencies. The congruency of the SEA subsystems is thus assured.

An assessment process based on a comparison with an equivalent full deterministic analysis is also applied, either to limited portions of the interstage hybrid models to verify the accuracy of the coupling loss factors among specific SEA and FEM subsystems, or to the entire interstage models for an overall check of the hybrid approach's accuracy. For each examined system, the results are compared in the frequency range for which the two models (hybrid and full FEM) are expected to be accurate.

4. Discussion

The results of the proposed integrated verification process are discussed in this section. Verification and validation are both crucial in assessing the project robustness. However, the analysis proposed here has been performed at the design level; thus the validation with experimental data is to be performed afterwards.

4.1. Resulting Pressure

In the framework of the VECEP, the Italian Aerospace Research Centre (*Centro Italiano Ricerche Aerospaziali*, CIRA) and the prime contractor, Avio SpA, are involved in the prediction of the aeroacoustic loads at lift-off due to the introduction of the first-stage solid rocket motor P120, a longer version of the currently used P80 motor. In this work, the empirical models used within the design verification process have been fed with the experimental data acquired using a 1/20 scaled mock-up of the VEGA launcher equipped with the first-stage solid rocket motor P80 [12,20]. Two experimental campaigns have been carried out by Avio, ELV and ONERA in the framework of the VEGA program [20,21] and by Avio, ELV and CIRA in the framework of the Italian Space Agency-funded project CAST (*Configurazioni AeroTermodinamiche per Sistemi di Trasporto spaziale*, AeroThermodynamic Configuration for space Transport Systems) [12,22].The launcher mock-up has been conceived to reproduce the same acoustic environment generated by the full-scale VEGA first-stage solid rocket motor P80 at different launcher altitudes, from 0 to 75 m, corresponding to the first 4 s of the VEGA ascent trajectory during lift-off.

The free-field results achieved with the Eldred standard model and the Eldred-based uncorrelated and correlated models have been compared with the VEGA and CAST experimental results for three launcher altitudes (0, 10 and 75 m) in [14]. The uncorrelated model has shown to provide a better prediction at 0 and 10 m, while the correlated model is better at 75 m, matching the experimental data with an uncertainty of about 5 dB.

The reason for the different matching between the correlated/uncorrelated models and the experiments' data can be explained reasonably with the different turbulent mechanisms that occur at different altitudes. At 75 m, the launcher is far from the launching pad, and, as a consequence, the jet does not interact with the ground. Indeed, the turbulence structures are still unchanged, and the correlation is preserved. Because the correlated model is developed to reproduce the jet turbulent scales' correlation, it is expected to be more suitable for reproducing the free jet conditions at the altitude of 75 m. Conversely, at the reduced distances of 0 and 10 m between the launcher and the launching pad, the jet turbulent structures are broken and mixed as a result of the interaction with the launching pad, and, as a consequence, the turbulent structures become more uncorrelated.

The effect of the launcher has thus been investigated with the hybrid approach. A three-dimensional multifrequency computation has been performed up to a maximum frequency of 1000 Hz, and the results have lastly been processed to achieve third-octave spectra. The uncorrelated model was used for 0 and 10 m, while the correlated model was used for 75 m. The output of this process was then passed as input to the hybrid procedure for analyzing the launcher response, as described in the following section.

The results of the BEM calculation are shown in Figure 4, where they are compared, in terms of SPL, with the standard Eldred results and the data from the two experimental campaigns.

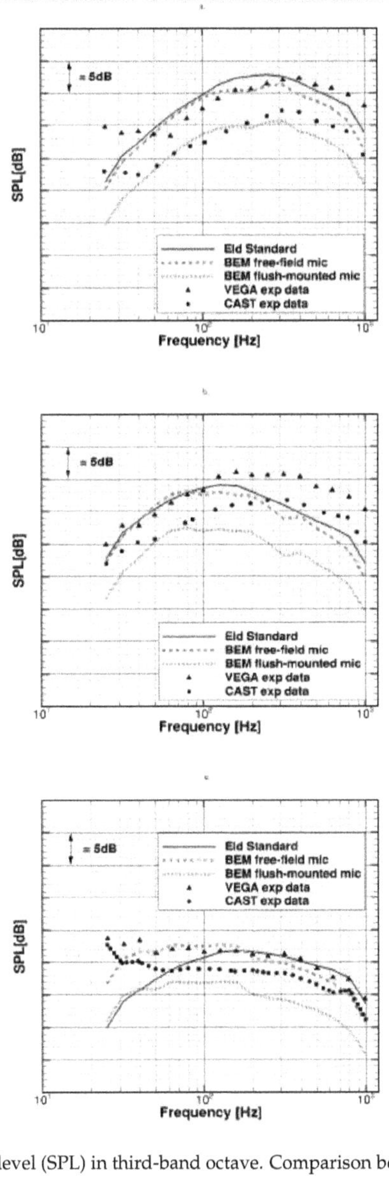

Figure 4. Sound pressure level (SPL) in third-band octave. Comparison between hybrid model results and experimental data collected at 0 (**a**), 10 (**b**) and 75 (**c**) m of altitude of the launcher interstages 2–3.

The SPL values have been computed in two sets of points, representing far- and near-field point. Far-field points, distant from the launcher by 15% of the nozzle diameter, have been dubbed in the figure as "free-field microphones". Near-field points, selected on the rocket surface itself, are indicated as "flush-mounted microphones". The gap observed between the free-field and the flush-mounted microphones resembles that occurring between the two experimental datasets, suggesting a different position of the microphone in the respective experimental set-up.

4.2. *Vibroacoustic Analysis*

The description of the hybrid model's response to the external acoustic field and internal acoustic field generated at lift-off by the P120 engine are reported, respectively, in terms of an average acceleration and a root-mean-square (RMS) pressure.

The acceleration data are reported as band-limited RMS spectrum responses strictly associated to the frequency band selected for the analysis. In the analysis, the damping was assumed as constant on the frequency, and similar was assumed for all the structural (values set to 0.04) and the acoustic (values set to 0.001) subsystems. The response of some equipment is reported in Figure 5 as an example.

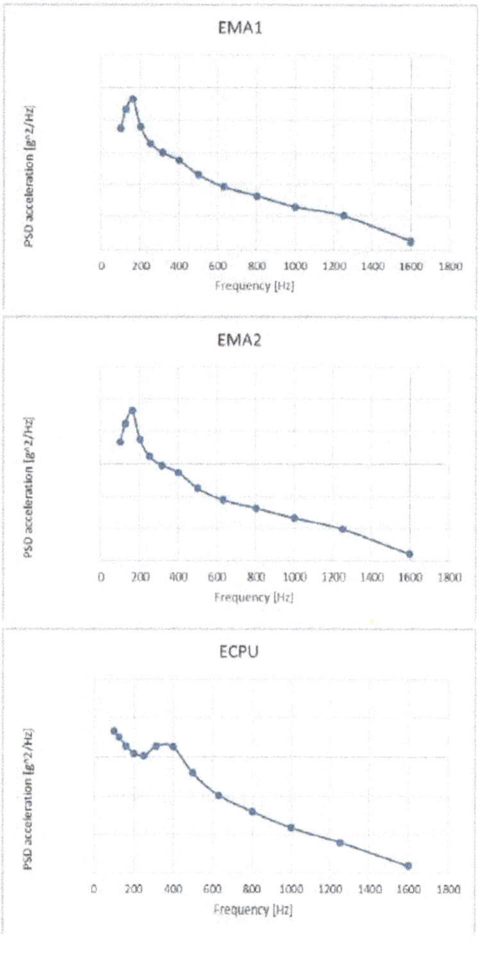

Figure 5. Accelerations calculated for some equipment in the IS01 interstage.

For IS01, the acceleration levels were very high in the low frequency range. In particular, the main contribution to the vibration energy was clearly associated to the structure's ring frequency, along with global modes' impact occurring at a lower frequency; above ring frequencies, the acceleration levels attenuated significantly for almost all the equipment, with some high-frequency contributions derived from the coincidence frequencies effect. In general, the dynamics of the lumped masses simulating the equipment appears as confined to the low–mid frequency.

The average pressure level associated to the IS01 cavity is reported in Figure 6, described in decibel values, referred to the band-limited RMS spectrum response, in which the reference pressure is the standard pressure value, 2×10^{-5} Pa. The responses reflect the same distribution against the frequency of the structural interstage, with the higher energy concentrated at the characteristic frequency already identified for the structure.

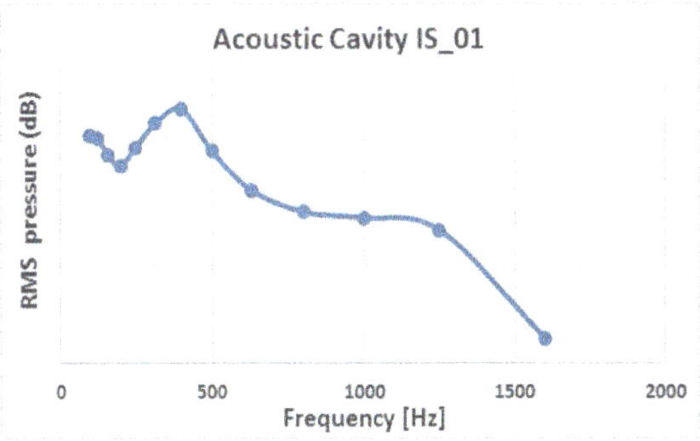

Figure 6. The acoustic cavity response in terms of the root-mean-square (RMS) pressure (dB) for the IS01 interstage.

5. Conclusions

An innovative, multi-domain verification process has been conceived and applied to the design of an enhanced version of the VEGA space launcher. The procedure has allowed a virtual, time- and cost-saving testing activity, developed by the prime contractor, Avio, in collaboration with CIRA SCpA.

The results are encouraging, and potentially the adoption of similar integrated and hybrid approaches can be foreseen in space and aviation industries.

Acknowledgments: The activity was supported and financed by Avio, who provided the interstage FEM models and all necessary data and information.

Author Contributions: All the authors have substantially contributed to the work. D.B., F.P.A. and P.V. also prepared the manuscript.

Conflicts of Interest: The authors declare no conflict of interest. The founding sponsors had no role in the design of the study; in the collection, analyses, or interpretation of data; or in the decision to publish the results. In the writing of the manuscript, the sponsors asked to hide the absolute references to data in Figures 2 and 4–6 because they represent an industrial secret and confidential information.

Abbreviations

The following abbreviations are used in this manuscript:

MDPI	Multidisciplinary Digital Publishing Institute
CFD	Computational Fluid Dynamics
VECEP	VEga Consolidation and Evolution Program
BEM	Boundary Element Method
FEM	Finite Element Method
SEA	Statistical Energy Analysis

References

1. European Cooperation for Space Standardization (ECSS). *Space Engineering Verification—ECSS-E-10-02A*; ECSS: Noordwijk, The Netherlands, 1998.
2. Citarella, R.; Federico, L.; Cicatiello, A. Modal acoustic transfer vector approach in a FEM-BEM vibro-acoustic analysis. *Eng. Anal. Bound. Elements* **2007**, *31*, 248–258.
3. Cherian, A.; George, P.G.; Prabha, C. Response analysis of payload fairing due to acoustic excitation. *Int. J. Sci. Technol. Res.* **2015**, *4*, 302–305.
4. Shorter, P.J.; Langley, R.S. Vibro-acoustic analysis of complex systems. *J. Sound Vib.* **2005**, *288*, 669–700.
5. Pirk, R.; Souto, C.A. Deterministic, hybrid and statistical vibro-acoustic models—A methodology to determine the VLS payload fairing acoustic behavior. *J. Aerosp. Technol. Manag.* **2014**, *10*, 1–9.
6. Barbarino, M. Aeroacoustics Methods for Low-Noise Technologies Design. Ph.D. Thesis, Universitá di Napoli Federico II, Napoli, Italy, 2013.
7. Barbarino, M.; Bianco, D. *BEM-FMM Simulation of Uniform Mean Flows with a New Internal-Point Algorithm for the CHIEF Spurious Solutions Removal*; ICSV: Wien, Austria, 2016.
8. Eldred, K.M. *Acoustics Loads Generated by the Propulsion System*; NASA SP-8072; NASA: Washington, DC, USA, 1971.
9. Candel, S. Analysis of the sound field radiated by ariane launch vehicle during lift-off. In *La Recerche Aerospatiale: Bulletin Bimestriel de L'Office Nationale d'Etudes et des Recherches Aerospatiales*; O.N.E.R.A.: Palaiseau, France, 1983; pp. 17–33.
10. Varnier, J.; Raguenet, W. Experimental characterization of the sound power radiated by impinging supersonic jets. *AIAA J.* **2002**, *40*, 825–831.
11. Varnier, J. Simplified approach of jet aerodynamics with a view to acoustics. *AIAA J.* **2006**, *44*, 1690–1694.
12. Casalino, D.; Barbarino, M.; Genito, M.; Ferrara, V. Hybridal empirical/computational aeroacoustic methodology for rocket noise modeling. *AIAA J.* **2009**, *47*, 1445–1460.
13. Maruf Morshed, M.; Hansen, C.H.; Zander, A. Prediction of acoustic loads on launch vehicle fairing during lift-off. *J. Spacecr. Rocket.* **2013**, *50*, 159–169.
14. Barbarino, M.; Adamo, F.P.; Bianco, D.; Bartoccini, D. Hybrid BEM/empirical approach for scattering of correlated sources in rocket noise prediction. *J. Sound Vib.* **2017**, *403*, 90–103.
15. Casalino, D.; Barbarino, M. Stochastic method for airfoil self-noise computation in frequency domain. *AIAA J.* **2011**, *49*, 2453–2469.
16. Di Francescantonio, P.; Hirsch, C.; Ferrante, P.; Isono, K. Side mirror noise with adapative spectral reconstruction. *SAE Tech. Pap.* **2015**, doi:10.4271/2015-01-2329.
17. Wang, H.; George, W.K. The integral scale in homogeneous isotropic turbulence. *J. Fluid Mech.* **2002**, *459*, 429–443.
18. ESIGroup. *VA One Users' Guide*; ESIGroup: Paris, France, 2009.
19. Timoshenko, S.P.; Goodier, J.N. *Theory of Elasticity*; McGraw-Hill: New York, NY, USA, 1951.
20. Gàly, D.; Elias, G.; Mascanzoni, F.; Foulon, H. Experimental acoustic characterization of the vega launch vehicle at lift-off. In Proceedings of the 6th International Symposium on Launchers Technologies, Centre National d'Etudes Spatiales, Munich, Germany, 8–11 November 2005.
21. Mascanzoni, F.; Contini, C. The acoustic environment of the vega lv at lift-off: The BEAT acoustic experimental test facility. In Proceedings of the 6th International Symposium on Launchers Technologies, Centre National d'Etudes Spatiales, Munich, Germany, 8–11 November 2005.
22. Casalino, D.; Barbarino, M.; Genito, M.; Ferrara, V. Rocket noise sources localization through a tailored beam-forming technique. *AIAA J.* **2012**, *50*, 2146–2159.

© 2018 by the authors. Licensee MDPI, Basel, Switzerland. This article is an open access article distributed under the terms and conditions of the Creative Commons Attribution (CC BY) license (http://creativecommons.org/licenses/by/4.0/).

applied sciences

MDPI

Article

Wavenumber-Frequency Analysis of Internal Aerodynamic Noise in Constriction-Expansion Pipe

Kuk-Su Kim [1,2], Ga-ram Ku [2], Song-June Lee [2], Sung-Gun Park [1] and Cheolung Cheong [2,*]

[1] Daewoo Shipbuilding and Marine Engineering, Geoje 53302, Korea; kuksu@dsme.co.kr (K.-S.K.); sgpark8@dsme.co.kr (S.-G.P.)

[2] School of Mechanical Engineering, Busan National University, Pusan 46241, Korea; garamku5671@pusan.ac.kr (G.-r.K.); sjlee@pusan.ac.kr (S.-J.L.)

* Correspondence: ccheong@pusan.ac.kr; Tel.: +82-051-510-2311

Received: 27 September 2017; Accepted: 2 November 2017; Published: 5 November 2017

Abstract: High-pressure gas is produced during the oil production process at offshore plants, and pressure relief devices, such as valves, are widely used to protect related systems from it. The high-pressure gas in the pipes connected to the flare head is burned at the flare stack, or, if it is nontoxic, is vented to the atmosphere. During this process, excessive noise is generated by the pressure relief valves that are used to quickly discharge the high-pressure gas to the atmosphere. This noise sometimes causes severe acoustic-induced vibration in the pipe wall. This study estimated the internal aerodynamic noise due to valve flow in a simple constriction-expansion pipe, by combining the large eddy simulation technique with a wavenumber-frequency analysis, which made it possible to decompose the fluctuating pressure into the incompressible hydrodynamic pressure and compressible acoustic pressure. First, the steady-state flow was numerically simulated, and the result was compared with a quasi-one-dimensional theoretical solution, which confirmed the validity of the current numerical method. Then, an unsteady simulation analysis was performed to predict the fluctuating pressure inside a pipe. Finally, the acoustic pressure modes in a pipe were extracted by applying the wavenumber-frequency transform to the total pressure field. The results showed that the acoustic pressure fluctuations in a pipe could be separated from the incompressible ones. This made it possible to obtain accurate information about the acoustic power, which could be used to assess the likelihood of a piping system failure due to acoustic-induced vibration, along with information about the acoustic power spectrum of each acoustic mode, which could be used to facilitate the systematic mitigation of the potential acoustic-induced vibration in piping systems.

Keywords: aerodynamic noise; valve noise; pipe flow; wavenumber-frequency analysis

1. Introduction

High-temperature and high-pressure gas is produced by various processes in chemical, petrochemical, and offshore oil production plants. Excessive noise and vibration can be generated when such gases are discharged, usually as they pass through a valve in the piping system used to protect the system and equipment from being damaged by such high-pressure gases. In the 1980s, pipe failures due to acoustic-induced vibration (AIV) were reported. Thus, the assessment and control of noise and vibration in piping systems have become essential parts of the design stage. Carucci et al. [1] first investigated pipe failure using nine pipe failure cases and 27 normal pipe cases. They suggested a criteria curve for the sound power and pipe diameter. This curve has been used as the basic screening method for the AIV of pipes. Eisinger [2] proposed simple piping design criteria by including the pipe wall thickness as an additional parameter. These were adopted in the Norsok Standard [3]. Recently, Bruce et al. [4] suggested new guidelines for the pipe AIV problem, in which the maximum vibration level is proportional to a nondimensional parameter that is given by the pipe diameter divided by the pipe wall thickness. The Energy Institute [5] published guidelines for the pipe AIV problem based on the likelihood of failure (LOF), which depends on parameters, such as the branch connection, main pipe diameter ratio, and material type.

However, all of these studies assumed that the main noise sources in a piping system are the pressure relief valves and estimated their acoustic power levels using an empirical formula that was based on API521 [6] and IEC 60534-8-3 [7]. In addition, they used an empirical criteria curve, the main parameters of which were the sound power and pipe dimensions obtained from the available pipe failure cases. Although these methods are very efficient at quickly screening a large number of pipes, it is difficult to apply efficient mitigation measures, because these tools rely on a limited number of input parameters to quantify the LOF, without any understanding of the relevant physical mechanisms. Recently, Agar et al. [8] proposed a finite element methodology based on fluid-structure coupling, which makes it possible to predict the dynamic stress in a complex piping system. However, even though this method could be used to quantitatively and effectively assess mitigation actions, the acoustic energy that was generated by a pressure-reducing device was estimated using a simple empirical formula, which inevitably limited its predictive accuracy with significant uncertainty.

In this study, to overcome this limitation, a systematic first-principles-based method was developed by combining high-order computational fluid dynamics (CFD) techniques with a wavenumber-frequency analysis. This method makes it possible to separate the acoustic pressure waves from the total pressure fluctuations in a pipe flow induced by valves, and to compute the acoustic power that is generated by the valves. First, the valve flow is predicted using high-resolution CFD techniques, in which the valve is modeled as a simple constriction-expansion pipe for simplicity. The validity of the calculated steady pressure is confirmed by comparing it to the theoretical solution obtained from quasi-one-dimensional (1D) inviscid compressible flow equations. Then, an unsteady simulation is performed to predict the fluctuating total pressure inside a pipe. Finally, the acoustic pressure modes in the pipe are extracted by applying the wavenumber-frequency transform to the total pressure field. The results show that the acoustic pressure fluctuations can be separated from the incompressible ones; this can provide accurate information about the acoustic power spectrum, which is used as an important parameter to determine the LOF of a piping system due to AIV.

The remainder of this paper is organized as follows. Section 2 introduces the theoretical background related to valve noise in piping systems, including the wavenumber-frequency analysis. Section 3 describes the governing equations and the numerical methods in detail. Section 4 presents the numerical results, including the steady and unsteady flow fields. Finally, the acoustic pressure modes in a pipe are extracted from the total pressure field using the wavenumber-frequency analysis, which makes it possible to predict the acoustic power spectrum. Section 5 concludes this paper.

2. Theoretical Background

2.1. Valve Flow and Its Quasi-1D Approximation

Commercially available valves may be categorized into globe valves, butterfly valves, pressure relief valves, and so on. A schematic diagram of a typical pressure relief valve is shown in Figure 1.

A relief valve is designed to open at a set pressure to protect a pressure vessel or other equipment. When the inlet pressure exceeds this set pressure, the relief valve opens and a portion of the gas is diverted through the valve system. As the fluid begins to move faster, its compressibility plays an increasingly important role, and it undergoes a significant change in density. If the pressure ratio across the constriction area exceeds a critical value, a choked flow with shock waves occurs [9], which causes additional aerodynamic noise.

The flow passage through a valve can be considered as a simple constriction and expansion pipe in terms of its cross-sectional area variation, whereby the area is reduced to a minimum at the throat of the valve and is then increased. An orifice plate works by essentially the same principle [10], in that it reduces the local pressure by providing a mechanical constriction in the pipe, although a valve is much more sophisticated than an orifice.

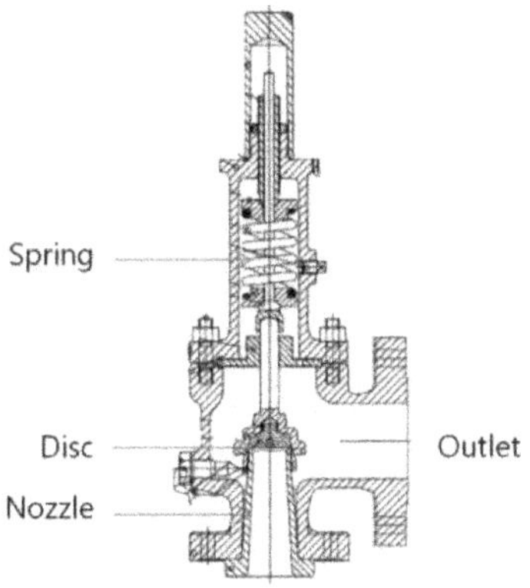

Figure 1. Schematic diagram of typical pressure relief valve.

The high-speed flow through a pipe with an orifice can be simplified as a quasi-1D compressible flow if the cross-sectional area is smoothly varied: the velocity components normal to the axial direction of the pipe (in the x and y directions) are small when compared to the velocity in the axial (z) direction, such that the flow field can be assumed to be a function of z only. In other words, it can be assumed that the flow is uniform across the cross-sectional plane of the pipe, which is called a quasi-1D flow. In this quasi one-dimensional flow, internal flow can be predicted analytically by solving the relevant equations presented by White [9].

As the pressure difference between the upstream and downstream regions increases, the mass flow rate also increases until the system reaches a critical condition, under which the mass flow no longer increases with an increase in the pressure difference. If the pressure difference increases further, the flow velocity in the throat remains constant at Mach 1. This is also called a choked condition [11]. When the outlet pressure is below the critical pressure, normal shock waves occur in the pipe, which are generated perpendicular to the flow direction. Generally, normal shock waves occur in a very thin region on the scale of 10^{-7} m, and are characterized by sudden changes in the flow quantities [12]. In a steady inviscid adiabatic 1D normal shock wave problem, the flow quantities downstream of the shock wave can be calculated from the upstream flow properties, which are used as the inlet conditions. The changes in the pressure and velocity across the shock wave are expressed as follows:

$$\frac{p_2}{p_1} = \frac{1}{\gamma+1}\left(2\gamma M_1^2 - (\gamma-1)\right) \tag{1}$$

$$M_2^2 = \frac{(\gamma-1)M_1^2+2}{2\gamma M_1^2 - (\gamma-1)} \tag{2}$$

where p_1 and p_2 are the pressures before and after shock wave, respectively. Because the shock wave is an adiabatic and irreversible process, an energy loss occurs and the entropy increases. The entropy variation across the shock wave is expressed as follows:

$$
\begin{aligned}
s_2 - s_1 &= c_p \ln \tfrac{T_2}{T_1} - R \ln \tfrac{p_2}{p_1} \\
&= c_p \ln \left[\left\{ 1 + \tfrac{2\gamma}{\gamma+1}\left(M_1^2 - 1\right) \right\} \frac{2 + (\gamma - 1)M_1^2}{(\gamma+1)M_1^2} \right] \\
&- R \ \ln \left\{ 1 + \tfrac{2\gamma}{\gamma+1}\left(M_1^2 - 1\right) \right\}
\end{aligned}
\tag{3}
$$

The entropy variation can be expressed using the total pressure change as follows:

$$
s_2 - s_1 = -R \ln \frac{p_{0,2}}{p_{0,1}} \tag{4}
$$

In the case of a gas with $T < 1000$ K, the specific heat can be assumed to be constant. A perfect gas with a specific heat constant is referred to as a calorically perfect gas. In this paper, the flow through a pipe can be assumed to be of a calorically perfect gas. Therefore, it can be assumed that the specific heat (c_p) and gas constant (R) are constant.

Because the mass flow rate through the pipe should be constant, the relationship between the cross-sectional area and Mach number in the pipe can be derived using the isentropic flow relation and mass conservation.

$$
\left(\frac{A}{A^*}\right)^2 = \frac{1}{M^2}\left\{\frac{2}{\gamma+1}\left(1 + \frac{\gamma-1}{2}M^2\right)\right\}^{\frac{\gamma+1}{\gamma-1}} \tag{5}
$$

where A is the area of the pipe and A^* is the critical area, which is the area of the throat. In this paper, because the geometry of the pipe is known, the theoretical solution derived from the above equations is compared with the numerical solution to confirm the validity of the numerical method.

2.2. Wavenumber-Frequency Analysis for Internal Aerodynamic Noise

AIV is known to originate with the high acoustic energy that is generated by pressure-reducing devices, which causes pipe shell vibration and thus induces excessive dynamic stress. Acoustic energy propagates at the speed of sound, i.e., the energy travels at the same speed as the compressible pressure waves. The total pressure fluctuation induced by valve flow in a pipe consists of compressible and incompressible pressure fluctuations. The incompressible part is the hydrodynamic pressure of a turbulent flow field with a relatively high energy content, whereas the compressible pressure fluctuation propagates at the speed of sound, which is determined by the compressibility of the fluid. The incompressible pressure fluctuation propagates at the convection speed of the mean flow. Therefore, it is essential to separate the compressible and incompressible acoustic pressure fluctuations to accurately assess the AIV. This can be achieved using a wavenumber-frequency analysis, which is described in detail below [13].

Because the cross-sectional shape of a pipe is circular, it is better to perform a wavenumber-frequency analysis in a cylindrical coordinate system. The analytical solution of the three-dimensional (3D) wave equation in a circular duct can be written in the following form:

$$
p(r,\theta,z,t) = \sum_{m=0}^{\infty}\sum_{n=0}^{\infty} J_m(k_{r,m,n}r)e^{jm\theta}e^{j\omega t} \times \left\{C_{+,m,n}e^{-jk^+_{z,m,n}z} + C_{-,m,n}e^{+jk^-_{z,m,n}z}\right\} \tag{6}
$$

where r, θ, and z are the coordinates in the radial, circumferential, and axial directions, respectively, and t is time. J_m is the 1st kind of Bessel function of the m-th order in the circumferential direction and ω is the angular frequency, k_z is the wavenumber in z-direction. C is the unknown constant and the subscript + and − mean the upstream and downstream parts, respectively.

Using the Fourier transform for the z-direction propagated wave, the spectral power density is expressed as follows:

$$
S(m,k_z,f) = \sum_{n=0}^{\infty}\frac{1}{J_m(k_{r,m,n}r_0)}\int_{t=-\infty}^{\infty}\int_{z=-\infty}^{\infty}\int_{\theta=0}^{2\pi} p(r_0,\theta,z,t)e^{-i(\omega t+m\theta-k_z z)}d\theta dz dt \tag{7}
$$

where $k_{r,m,n}$ is the radial wavenumber of the m-th circumferential mode and n-th radial mode, and k_z is the wavenumber in the axial direction. The dispersion relation of the internal acoustic waves to a constant mean flow of Mach number M can be expressed in the following form:

$$k_{z,m,n}^2 + k_{r,m,n}^2 = (k_0 + Mk_{z,m,n})^2 \tag{8}$$

Thus, the axial wavenumber in the positive and negative axial directions can expressed as follows:

$$k_{z,m,n}^{\pm} = \frac{\mp Mk_0 + \left[k_0^2 - (1 - M^2)k_{r,m,n}^2\right]^{1/2}}{1 - M^2} \tag{9}$$

Because the acoustic modes affecting the pipe wall vibration are the θ and z modes, the Fourier transform is considered only in the θ and z directions, whereas the coordinate in the r direction is treated as a constant. For the cut-on higher-order mode, the following relation must be satisfied.

$$k_0^2 - \left(1 - M^2\right)k_{r,m,n}^2 > 0 \tag{10}$$

In this paper, the discrete Fourier transform is performed using the Hanning window, which can be written in the following form:

$$S(m, p\Delta k_z, o\Delta f) = \frac{1}{N_t N_z N_\theta} \sum_{n=0}^{\infty} \frac{1}{J_m(k_{r_0,m,n} r_0)} \sum_{k=1}^{N_t\ 1} \sum_{l=0}^{N_z\ 1} \sum_{j=1}^{N_\theta\ 1} w_{klj} p_{klj} e^{2\pi i\left(\frac{ko}{N_t} + \frac{m\theta}{N_\theta}\ \frac{lp}{N_z}\right)} \tag{11}$$

$$w_{klj} = \frac{1}{2}\left[1 - \cos\left(\frac{2\pi\sqrt{j^2 + l^2 + k^2}}{\sqrt{(N_\theta - 1)^2 + (N_z - 1)^2 + (N_t - 1)^2}}\right)\right] \text{ : Hanning window}$$

2.3. Acoustic Power Level

The formula for the acoustic power of a fluid in a pipe is briefly introduced, because most of the existing empirical approaches consider the acoustic power level to be a critical parameter. The acoustic particle velocity in a duct with a circular cross-section can be described as follows [14]:

$$u_z(r, \theta, z, t) = \sum_{m=0}^{\infty} \sum_{n=0}^{\infty} \frac{k^+{}_{z,m,n}}{k_0 \rho_0 c_0} J_m(k_{r,m,n} r) \cos(m\theta)\, C_{+,m,n} e^{i(\omega t - k_{z,m,n}^+ z)} \tag{12}$$

From Equations (6) and (12), the acoustic power is defined as follows:

$$dW = I_z\, dA = p u_z dA$$
$$W = \int \overline{I}\, dA \quad where \quad \overline{I} = \frac{1}{T}\int_0^T I_z dt \tag{13}$$

where I_z is the intensity of the propagated acoustic wave, A is the cross-sectional area, and $\overline{I}$ is the time-averaged acoustic intensity.

After some algebra, the formula for the acoustic power can be written as follows:

$$W = \sum_{m=0}^{\infty} \sum_{n=0}^{\infty} \frac{k^+{}_{z,m,n}}{k_0 \rho_0 c_0} C_{+,m,n}^2 \frac{1}{2} \int_0^{2\pi} \cos^2(m\theta) d\theta \frac{R^2}{2} \left\{ J_m(k_{r,m,n} R)^2 - J_{m-1}(k_{r,m,n} R) J_{m+1}(k_{r,m,n} R) \right\} \tag{14}$$

Note that this is valid when the effects of the mean flow are negligible.

3. Governing Equations and Numerical Methods

3.1. Governing Equations

As the governing equations for the fluid flow in a pipe, the compressible Navier-Stokes equations are used in the following form:

$$\frac{\partial \rho}{\partial t} + \nabla \cdot \left(\rho \vec{v}\right) = 0 \tag{15}$$

$$\frac{\partial}{\partial t}\left(\rho \vec{v}\right) + \nabla \cdot \left(\rho \vec{v} \vec{v}\right) = -\nabla p + \nabla \cdot \overline{\overline{\tau}} + \vec{F} \tag{16}$$

where ρ is the fluid density, $\vec{v}$ is a velocity vector, p is the static pressure, $\overline{\overline{\tau}}$ is the viscous stress tensor, and $\vec{F}$ is the external body force. The stress tensor, $\overline{\overline{\tau}}$, is found as follows:

$$\overline{\overline{\tau}} = \mu\left[\left(\nabla \vec{v} + \nabla \vec{v}^{T}\right) - \frac{2}{3}\nabla \cdot \vec{v} I\right] \tag{17}$$

where μ is the dynamic viscosity, and I is the unit tensor. The energy conservation is described by the following equation:

$$\frac{\partial}{\partial t}(\rho E) + \nabla \cdot \left(\vec{v}(\rho E + p)\right) = -\nabla \cdot \left(\sum h_j J_j\right) + S_h \tag{18}$$

where E is the energy per unit mass, h_j is the heat transfer coefficient, and J_j and S_h are the diffusion flux and heat source, respectively.

3.2. Numerical Methods

First, a steady-state simulation was performed. The Reynolds-averaged Navier-Stokes equations (RANS) were used as the governing equations in the following form:

$$\begin{aligned} &\frac{\partial \rho}{\partial t} + \frac{\partial}{\partial x_i}(\rho u_i) = 0 \\ &\frac{\partial}{\partial t}(\rho u_i) + \frac{\partial}{\partial x_j}(\rho u_i u_j) = -\frac{\partial p}{\partial x_i} + \frac{\partial}{\partial x_j}\left[\mu\left(\frac{\partial u_i}{\partial x_j} + \frac{\partial u_j}{\partial x_i} - \frac{2}{3}\delta_{ij}\frac{\partial u_l}{\partial x_l}\right)\right] + \frac{\partial}{\partial x_j}\left(-\rho \overline{u_i' u_j'}\right) \end{aligned} \tag{19}$$

Equation (19) could be solved using the k–ω SST turbulence model, where k represents the turbulent kinetic energy, and ω denotes the specific turbulent dissipation rate. The k–ω SST model uses the k–ω model near the wall and a k–ε model in the region far from the boundary for accurate and stable calculations. This turbulence model is known to be the most suitable model for a supersonic internal flow16. The equations are written in the following form:

$$\frac{\partial}{\partial t}(\rho k) + \frac{\partial}{\partial x_i}(\rho k u_i) = \frac{\partial}{\partial x_j}\left(\Gamma_k \frac{\partial k}{\partial x_j}\right) + \widetilde{G}_k - Y_k + S_k \tag{20}$$

$$\frac{\partial}{\partial t}(\rho \omega) + \frac{\partial}{\partial x_i}(\rho \omega u_i) = \frac{\partial}{\partial x_j}\left(\Gamma_\omega \frac{\partial \omega}{\partial x_j}\right) + G_\omega - Y_\omega + D_\omega + S_\omega \tag{21}$$

where $\widetilde{G}_k$ is the production term for the turbulent kinetic energy k due to the velocity gradient, and G_ω is the generation term for the specific dissipation rate ω, Γ and Y are the effective diffusivity and dissipation terms, respectively. The k–ω SST model requires a cross-diffusion term D_ω, which contributes to the conversion relationship between the k–ω model and the k–ε model. S_k and S_ω are source terms. The density-based implicit solver was used to solve the RANS equation numerically. The flow was discretized using a second-order upwind scheme, and the turbulent kinetic energy and dissipation rate were found using a first-order upwind scheme.

Then, the large eddy simulation (LES) method with a Smagorinsky-Lilly model was used for the unsteady simulation to predict the fluctuating pressure waves inside a pipe with a high resolution. The turbulent viscosity of the Smagorinsky-Lilly model was calculated as follows.

$$\mu_t = \rho L_s^2 \left|\overline{S}\right| \; where \; \left|\overline{S}\right| = \sqrt{2\overline{S}_{ij}\overline{S}_{ij}} \tag{22}$$

where $\overline{S}_{ij}$ and L_s are the rate of the strain tensor and the mixing length, respectively. These are expressed in the following form:

$$\overline{S}_{ij} = \frac{1}{2}\left(\frac{\partial \overline{u}_i}{\partial x_j} + \frac{\partial \overline{u}_j}{\partial x_i}\right) \tag{23}$$

$$L_s = \min(\kappa\, d, C_s\Delta) \tag{24}$$

where κ is the von Karman constant, d is the distance to the nearest wall, C_s is the Smagorinsky constant, and Δ is the local grid size, which is defined as follows:

$$\Delta = V^{1/3} \tag{25}$$

Although Lilly proposed C_s = 0.23 for a homogeneous isotropic turbulence, because it is known that excessive damping occurs for large pressure fluctuation components, $C_s \approx 0.1$ was used to obtain an accurate unsteady pressure in the current numerical simulation. The second-order upwind scheme was used as the spatial discretization scheme. The maximum Courant number was set to 1 and the time step was 0.05 ms, for which the Nyquist frequency was 10,000 Hz. The nonreflective boundary was applied to the outlet to accurately capture the acoustic pressure waves in the computation domain, by eliminating any possible contamination due to reflected waves from the outflow boundary. The simulations were performed with the commercial CFD code Fluent, which is based on finite volume methods. A wavenumber-frequency analysis was performed and the results were combined with the LES results to separate the compressible acoustic pressure waves from the incompressible hydrodynamic pressure fluctuations.

4. Numerical Results

First, the steady-state results were used to confirm the validity of the current numerical method through a comparison of its prediction result with those of the analytical solution, and then the unsteady LES results were investigated. The total pressure and temperature values were prescribed on the inlet and outlet surfaces of the pipe as boundary conditions. The boundary values of the pressure and temperature were determined based on the design values of the piping system that were obtained using the software HYSYS, which is widely used for process simulation.

4.1. Axisymmetric Steady Simulation

The valve flow was simplified as a quasi-1D flow in a pipe consisting of uniform inlet and outlet ducts with the same diameter (0.292 m), and an orifice with a minimum diameter of 0.144 m, as shown in Figure 2. These dimensions were based on the data for a real valve in a pipe that is used in an existing ocean plant.

The inlet and outlet flow conditions were prescribed using the following design data. The inlet total pressure was 9.0 bar, inlet static pressure was 7.0 bar, inlet temperature was 523 K, outlet pressure was 4.4 bar, and outlet temperature was 433 K. An axisymmetric boundary condition was applied along the center line of the pipe, and the no-slip boundary condition was applied on the inner pipe wall. The number of grid points in a typical calculation was 200,000.

The results of the axisymmetric steady simulation were compared with those that were obtained from the quasi-1D theory. Figure 3 shows the convergence time of the simulation in terms of the inlet and outlet mass flows. The mass flow rates are seen to converge quickly after 5000 iterations to the same value (25.7 kg/s) at the inlet and outlet. Because the flow is an inflow at the inlet, the mass flow rate has a positive value, and vice versa at the outlet. All of the subsequent analyses were performed using the converged numerical results.

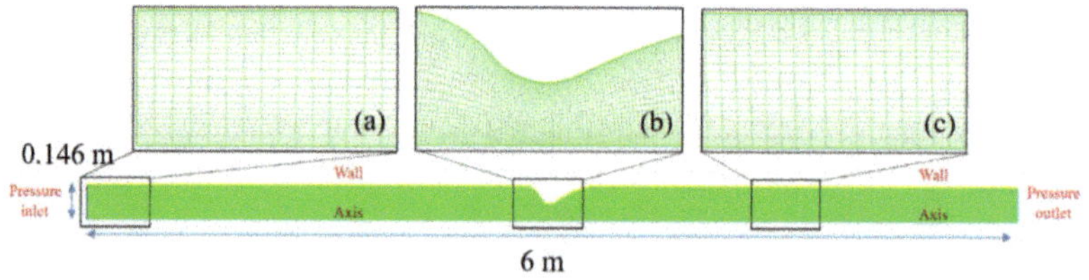

Figure 2. Grids in physical domain for pipe flow for steady and unsteady computational fluid dynamics (CFD) simulations (**a**) pipe inlet; (**b**) pipe throat; (**c**) pipe outlet.

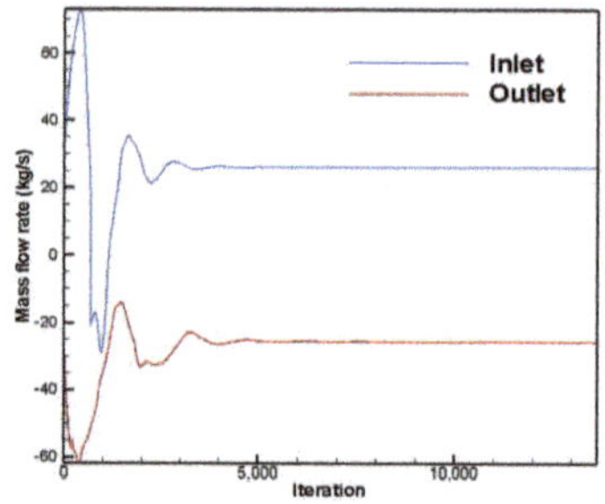

Figure 3. Time histories of mass flow rates.

Figure 4 shows the iso-contours of the static pressure, temperature, and velocity magnitude in the vicinity of the orifice. Their variations on the center line (y = 0) of the pipe are also shown in Figure 5, where the velocity results are described in terms of the Mach number. Note that the orifice is located at x = 1 m. Upstream from the orifice, the total pressure is approximately 9.0 bar, and it continuously decreases as the flow approaches the orifice. Downstream of the orifice, the static pressure increases sharply up to 4.4 bar and remains almost constant far downstream. It is found that the shock wave is located at approximately x = 1.1 m. Other variables show similar behaviors. To verify the numerical results, these distributions along the center line were compared with those of the theoretical solutions. To improve the understanding of the numerical solutions, three conditions for the numerical results were compared with the theoretical ones: steady with friction, steady without friction, and unsteady with friction. Because the analytical solution was based on the 1D inviscid compressible Euler equations, the numerical results without viscous friction showed the closest agreement with the theoretical ones. The steady and unsteady RANS results showed similar behaviors, except for the oscillation of the steady results in the vicinity of the shock waves.

However, there was excellent agreement between the numerical and theoretical results, except for the region just after the shock wave, where the complexity of the interaction between the shock wave and boundary layer invalidated the 1D approximation based on the theoretical approach. These results ensured the validity of the current numerical method.

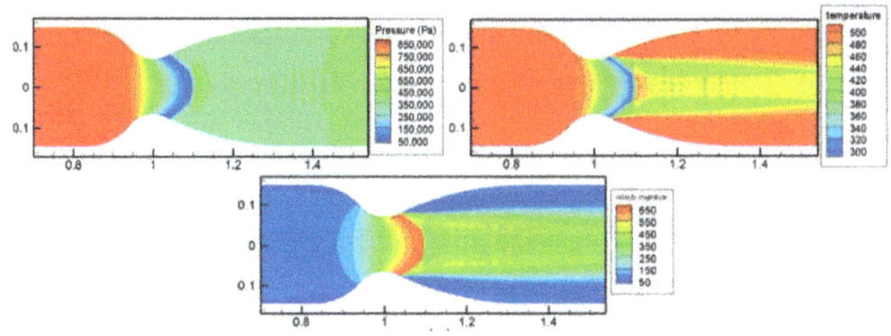

Figure 4. Converged results: (**a**) static pressure [Pa]; (**b**) temperature [K]; and, (**c**) velocity magnitude [m/s].

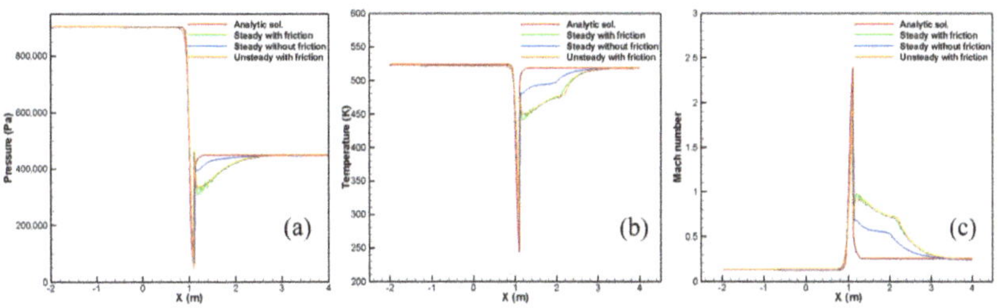

Figure 5. Spatial distributions of (**a**) static pressure; (**b**) temperature; and, (**c**) Mach number along axis of pipe.

4.2. Three-Dimensional Steady Simulation

To obtain the initial values for the unsteady LES simulation, a 3D steady RANS simulation was performed using the same inlet and outlet boundary conditions as the axisymmetric simulation. The geometry and grids are shown in Figure 6. The number of grids used was 2,700,000. To obtain more reliable unsteady results by reducing the possible reflection waves from the outlet, the length of the computational domain was extended so that the outlet was 20.5D away from the orifice in the downstream direction.

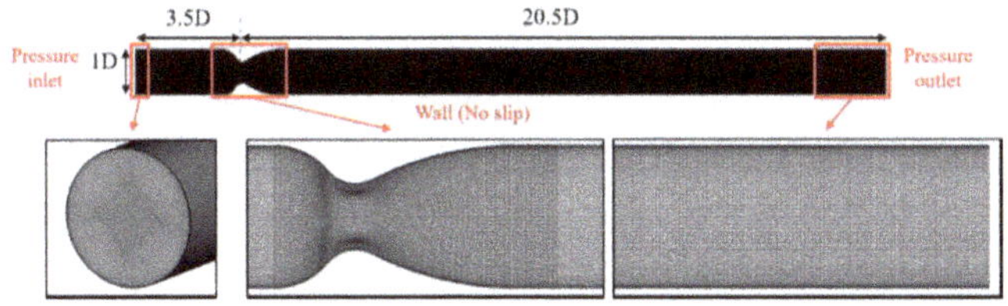

Figure 6. Grids in physical domain for pipe flow for three-dimensional (3D) steady CFD simulations with applied boundary conditions.

To verify the results, the mass flow rate predicted is shown in Figure 7. It can be seen that the mass flow rate converges at the inlet and outlet. The iso-contour distributions of the pressure, temperature, and Mach number on the central cross-sectional plane in the vicinity of the orifice are shown in Figure 8. These results are almost the same as those in the axisymmetric case. The spatial variations of these variables along the pipe axis are shown in Figure 9. Again, there is an excellent agreement between the numerical and theoretical results, with the exception of the transient region.

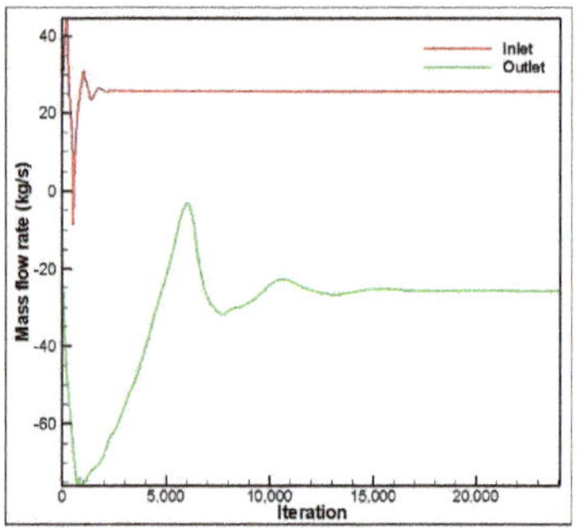

Figure 7. Time histories of mass flow rates.

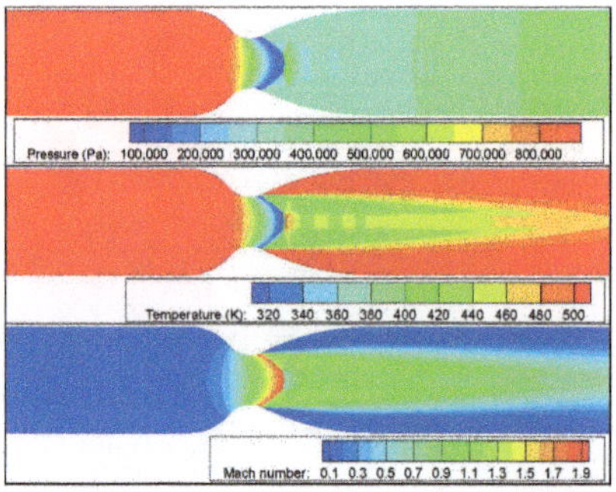

Figure 8. Flow fields for 3D Reynolds-averaged Navier-Stokes equations (RANS): static pressure [Pa], temperature [K], and velocity [m/s].

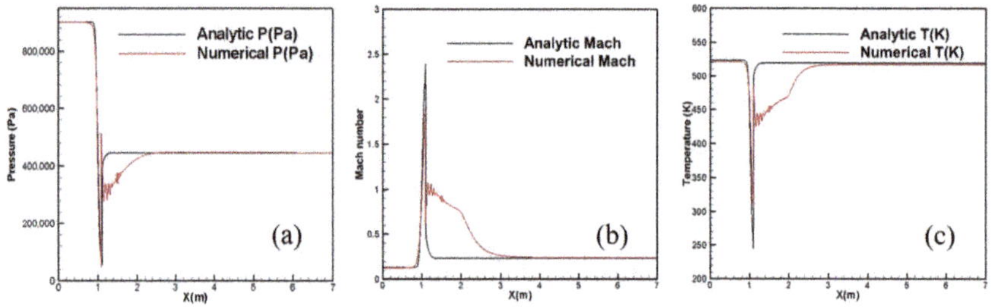

Figure 9. Spatial distributions of properties along pipe axis: (**a**) static pressure; (**b**) Mach number; (**c**) temperature.

As shown in Figure 10, a complicated flow structure is formed in the region downstream of the shock wave as a result of the complex interaction between the shock wave and the boundary layer at the wall surface. These complex interactions generate a strong turbulent flow, and thus cause aerodynamic noise. The following section shows how this was investigated in detail.

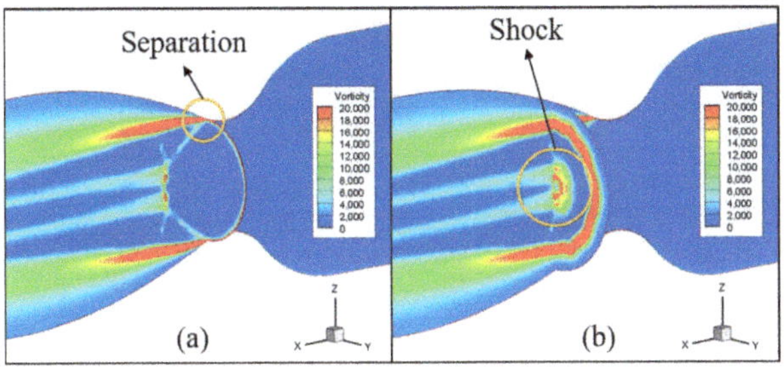

Figure 10. Vorticity contours near throat: (**a**) separation location due to shock-boundary layer interaction; (**b**) shock location at center line.

4.3. Three-Dimensional Unsteady Simulation

This section discusses an unsteady simulation that was performed using the LES to capture the compressible acoustic field, as well as the incompressible hydrodynamic flow field. For efficient computation, the 3D RANS steady simulation result was used as an initial condition for the LES simulation. The unsteady simulation was carried out with a time step of 0.05 ms and 8,220,000 grids. In the case of the unsteady LES analysis, the unsteady flow characteristics should be captured by the numerical method combined with its mesh. The important unsteady phenomena in the CAA simulation are two-fold: one is aerodynamic noise source mechanism and the other is acoustic wave propagation. Therefore, the quality of grid for CAA simulation is generally determined by the highest frequency of the concerned frequency range of the problem. In other words, as the grid-resolution increases, the resolved frequency range also increases. An important aerodynamic noise source mechanism in the current problem is closely related to the turbulence in the boundary layer, especially, downstream of the orifice. The important length scale related to the aeroacoustic source is $y+$ so that the minimum grid length is determined, such that $y+ = 10$. The length scale related to acoustic wave propagation is the wavelength of the highest frequency component, which is set to be 10,000 Hz in the current problem. The maximum grid length is matched to be $\Delta x_{max} = \lambda_{min}/8$, where λ_{min} is the wavelength of the acoustic wave of 10,000 Hz. The geometry and boundary conditions were the same as those for the 3D RANS simulation, as shown in Figure 6.

Figure 11 shows the variations in the mass flow rates at the inlet and outlet in the LES simulations. Because the 3D RANS result was used for the initial condition, a transient period is seen before the converged numerical solutions occur after 1500 iterations. Figure 12 shows the distributions of the static pressure, turbulent viscosity, and vorticity on the central cross-sectional plane and in an enlarged area downstream of the orifice. The strong turbulent fluctuation due to the complex interaction between the shock and the separated boundary layer is observed in the region just downstream of the shock wave.

There are two types of excitation sources for the vibration of a pipe wall: the compressible pressure fluctuation that is associated with acoustic waves, and the incompressible pressure fluctuation, which is associated with a turbulent flow. As shown in Figure 12c, the incompressible pressure fluctuations associated with a strong vortex convected with the mean flow can induce pipe vibration. This phenomenon is called flow-induced vibration. In addition, the acoustic waves generated by the turbulent fluctuation and its interaction with the solid surface propagate at the velocity of sound plus the mean flow velocity. The compressible pressure fluctuation associated with the acoustic waves also causes pipe vibration, which is called AIV. However, because the compressible pressure fluctuation propagates at the speed of sound, it is difficult to directly observe it from the flow field. Therefore, the compressible pressure fluctuation and incompressible pressure fluctuation were separated using a wavenumber-frequency analysis.

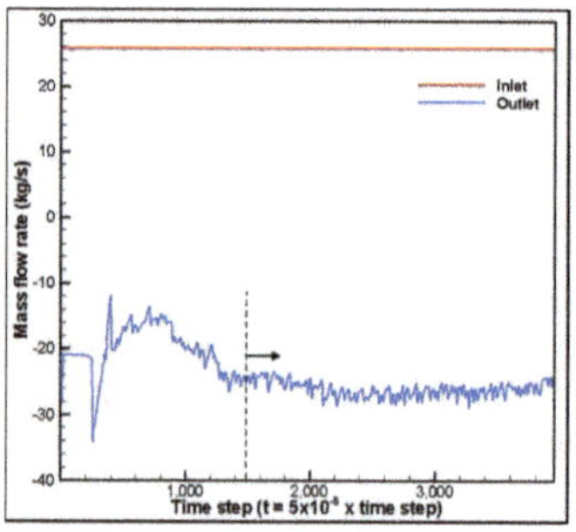

Figure 11. Time histories of mass flow rates for large eddy simulation (LES).

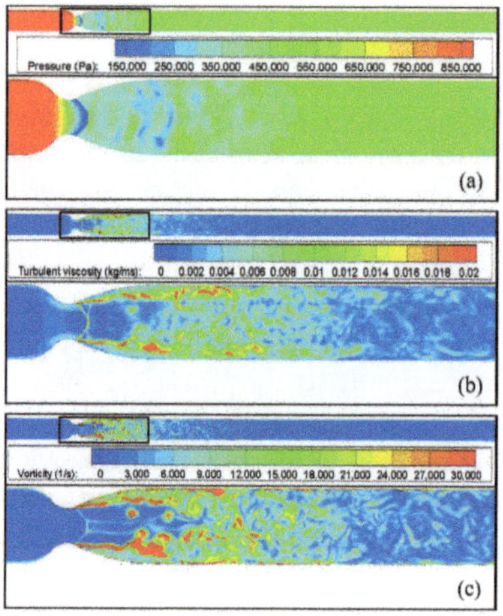

Figure 12. Flow fields for LES: (**a**) static pressure [Pa]; (**b**) turbulent viscosity [kg/ms]; and, (**c**) vorticity [1/s].

4.4. Wavenumber-Frequency Analysis

The wavenumber-frequency analysis was performed using the LES results presented in Section 4.3. Figure 13 shows the wavenumber-frequency transform of the static pressure on a cylindrical inner duct wall, i.e., $r = d/2$ for a length of 1 m from the pipe outlet. The frequency and time intervals are $\Delta f = 10$ Hz and $\Delta t = 5 \times 10^{-5}$ s, respectively. The spatial and wavenumber intervals are $\Delta z = 3.844 \times 10^{-3}$ m, $\Delta\theta = 10°$, and $\Delta k_z = 1.004$, respectively. The average speed of sound and mean flow Mach number over the target region are $c_0 = 317$ m/s and $M = 0.296$, respectively. The speed of sound is considered to be the smallest value in the entire flow field.

The black cut-off lines corresponding to the modes of $(m, n) = (0{:}4, 0{:}3)$ are also drawn to make it easier to understand the behavior of the pressure waves inside the pipe in the wavenumber-frequency domains. The m-mode is the circumferential mode, and the n-mode is the radial mode. The dashed line represents the incompressible pressure wave convecting at a constant mean flow speed of $0.88U_0$. The strong compressible acoustic waves propagating at speeds of $c + U_0$ (downstream propagating waves) and $-c + U_0$ (upstream propagating waves) can be identified in the wavenumber-frequency diagram of the $m = 0$ and $n = 0$ modes. The condition for the higher-order modes is shown in Equation (10). From Equation (9), the variation in the z-direction wavenumber, k_z, for the higher-order modes is described in Figure 13 in the parabola form, according to the n-modes [14]. It can be seen from

Figure 13 that the $n = 0$ and $n = 1$ modes contributes dominantly, whereas the higher-order modes with $n = 2$ or higher are relatively negligible. Cowling mentioned that pipe failure due to acoustic waves occurs in the range of 500–2000 Hz [8,15]. In this study, the higher-order modes above $n = 2$ occurred at frequencies greater than 2000 Hz. Thus, they did not belong to the pipeline failure region due to acoustic waves. The line of the $n = 0$ mode shows the velocity of the acoustic wave propagating upstream and downstream, where the area inside the line is the compressible pressure fluctuation component, and the area outside the line is the incompressible pressure fluctuation component. Based on the results of the wavenumber-frequency analysis, it can be seen that the compressible and incompressible pressure fluctuations can be separated from the entire flow field, as shown in Figure 14.

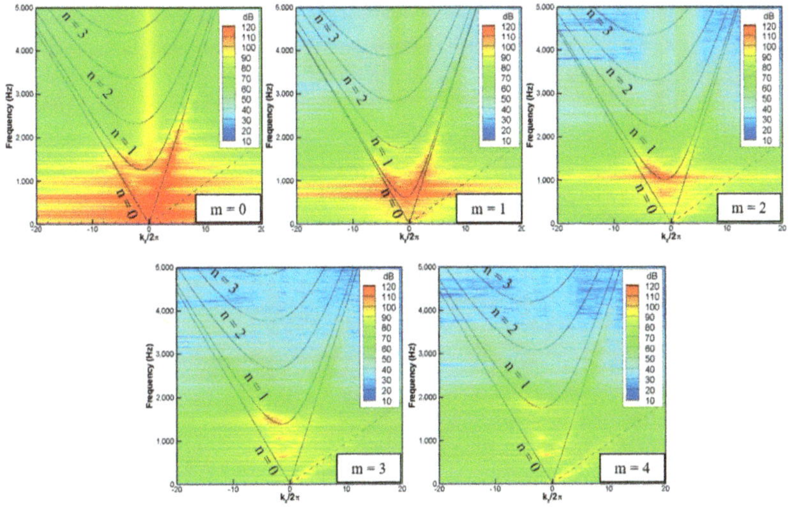

Figure 13. Wavenumber-frequency diagram of surface static pressure on cylindrical plane $r = d/2$.

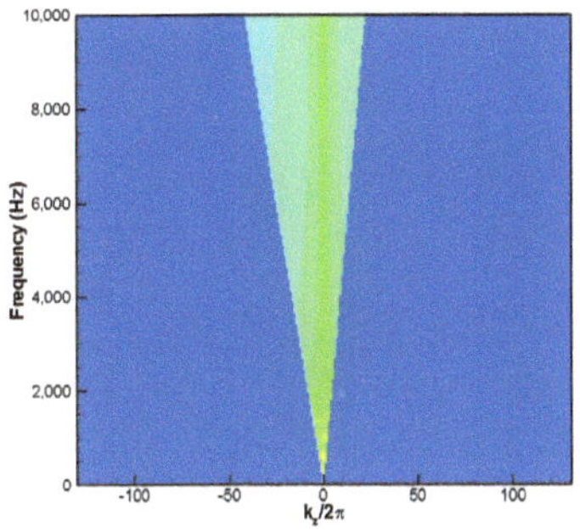

Figure 14. Region of compressible pressure fluctuation component for $m = 0$ mode.

Figure 15 compares the acoustic power level spectrum that was obtained from the inverse wavenumber-frequency transform of the compressible and total pressure fluctuation components. It shows that the acoustic power spectrum can be separated from that due to the total pressure fluctuation. Figure 16 shows the acoustic power spectrum of each circumferential mode. These were obtained from the inverse wavenumber-frequency transform of the wavenumber-frequency diagram, as shown in Figure 14. The contribution of each circumferential acoustic mode to the total acoustic power can be quantitatively assessed based on these results. This information can be utilized because the higher modes ($m > 0$) are known to be more critical in causing flow-induced vibration in a pipe. Carucci et al. [1,4] presented an empirical formula for predicting the acoustic power level in pipes based on flow field information. Eisinger et al., Riegel et al. and CSTI proposed standard values for the acoustic power level based on the pipe dimensions [4], and recommended designing a pipe with a value smaller than the standard value. Table 1 compares the overall acoustic power levels that are predicted using the current approach with those obtained using these empirical methods. Note that the empirical formulas cannot consider the influence of the mean flow, because these models were developed using a jet noise model that does not include the mean flow effects. Therefore, in this study, the acoustic power was calculated without considering the influence of the mean flow for a fair comparison. It can be seen that the empirical approaches over-estimate the acoustic power levels. The reason for this is that the empirical formula were developed using the data obtained from the real valves with more complex geometries than the present model. However, the empirical methods cannot provide detailed information about the spectral distribution of acoustic power levels that the present method can do.

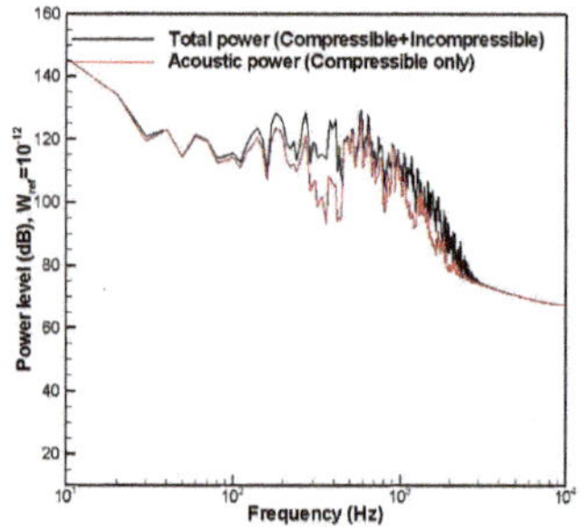

Figure 15. Power level spectrum of total and acoustic component.

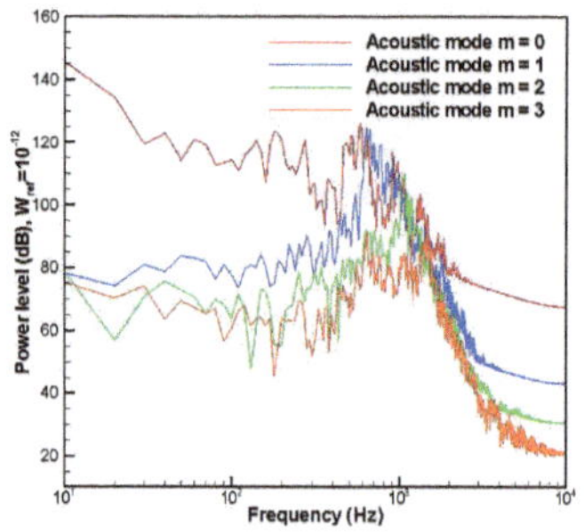

Figure 16. Power level spectrum of each mode.

Table 1. Overall acoustic power levels found using empirical formulas.

Acoustic Power Level, L_w (dB)	
Carucci et al. (1982)	163.24
Eisinger et al. (1996)	167.85
Riegel et al. (2007)	169.69
CSTI (2013)	169.85
Present study	146.32

5. Conclusions

The LES technique and a wavenumber-frequency analysis were combined to estimate the internal compressible acoustic waves and incompressible hydrodynamic pressure waves in a simple constriction-expansion pipe. First, the steady-state flow was numerically simulated, and the result was compared with a quasi-1D theoretical solution, which confirmed the validity of the current numerical method. Then, an unsteady simulation was performed to predict the unsteady pressure fluctuation inside a pipe. Next, the compressible acoustic pressure waves and incompressible hydrodynamic pressure waves in the pipe were separated by applying a wavenumber-frequency transform to the total pressure field. Finally, the inverse wavenumber-frequency transform facilitated the estimation of the acoustic power spectrum by separating the compressible and incompressible pressure fluctuations. The estimated acoustic power level was compared with those that were obtained using the existing empirical models, which highlighted the fact that the current method can provide more detailed information about the acoustic source for the so-called AIV of a piping system, and thus facilitate systematic measures for the mitigation of this AIV.

Acknowledgments: This research was supported by the Basic Science Research Program through the National Research Foundation of South Korea (NRF), funded by the Ministry of Education (NRF-2016R1D1A1A099 18456).

Author Contributions: Cheolung Cheong provided the overall numerical analysis schemes. Kuk-Su Kim and Ga-ram Ku carried out the numerical simulations and worked on the analysis of numerical results. Song-June Lee developed the code of wavenumber-frequency analysis. Sung-Gun Park provided the operation conditions of the target problem and the empirical formulas available.

Conflicts of Interest: The authors declare no conflict of interest.

References

1. Carucci, V.A.; Mueller, R.T. *Acoustically Induced Piping Vibration in High Capacity Pressure Reducing Systems*; ASME 82-WA/PVP-8; The American Society of Mechanical Engineers: New York, NY, USA, 1982.
2. Eisinger, F.L. *Designing Piping Systems against Acoustically-Induced Structural Fatigue*; ASME 1996, PVP-328; The American Society of Mechanical Engineers: New York, NY, USA, 1996.
3. Veritas, D.N. *L-002 Piping System Layout, Design and Structural Analysis*, 3rd ed.; NORSOK Standard: Lysaker, Norway, 2009.
4. Bruce, R.D.; Bommer, A.S.; LePage, T.E. Solving acoustic-induced vibration problems in the design stage. *Sound Vib.* **2013**, *47*, 8–11.
5. Energy Institute. *Guidelines for the Avoidance of Vibration Induced Fatigue Failure in Process Pipework*, 2nd ed.; Energy Institute: London, UK, 2008.
6. API Standard 521. *Guide for Pressure Relieving and Depressuring System*, 4th ed.; American Petroleum Institute (API): Washington, DC, USA, 1997.
7. IEC 60534-8-3. *Noise Considerations—Control Valve Aerodynamic Noise Prediction Method*; The International Electrotechnical Commision (IEC): Geneva, Switzerland, 2010.
8. Agar, M.; Ancian, L. Acoustic-Induced Vibration: A New Methodology for Improved Piping Design Practice. In Proceedings of the Offshore Technology Conference Asia, Kuala Lumpur, Malaysia, 22–25 March 2016.
9. White, F.M. *Fluid Mechanics*, 6th ed.; McGraw-Hill: New York, NY, USA, 2009.
10. Blake, W.K. *Mechanics of Flow-Induced Sound and Vibration Volume 1: General Concepts and Elementary Source*; Academic Press: New York, NY, USA, 1986.
11. Park, K.A.; Choi, Y.M.; Choi, H.M.; Cha, T.S.; Yoon, B.H. The evaluation of critical pressure ratios of sonic nozzles at low Reynolds numbers. *Flow Meas. Instrum.* **2001**, *12*, 37–41. [CrossRef]
12. Anderson, J.D., Jr. *Fundamentals of Aerodynamics*; Tata McGraw-Hill Education: New York, NY, USA, 2010; pp. 543–573.
13. Van Herpe, F.; Baresh, D.; Lafon, P.; Bordji, M. Wavenumber-frequency analysis of the wall pressure fluctuations in the wake of a car side mirror. In Proceedings of the 17th AIAA/CEAS Aeroacoustics Conference, Portland, OR, USA, 5–8 June 2011; pp. 3823–3839.

14. Munjal, M.L. *Acoustics of Ducts and Mufflers*, 2nd ed.; John Wiley & Sons: Chichester, UK, 2014; pp. 2–20.
15. Cowling, J. *Acoustic and Turbulence/Flow Induced Vibration in Piping Systems: A Real Problem for LNG Facilities*; Perth Convention and Exhibition Centre: Houston, TX, USA, 2016.

© 2017 by the authors. Licensee MDPI, Basel, Switzerland. This article is an open access article distributed under the terms and conditions of the Creative Commons Attribution (CC BY) license (http://creativecommons.org/licenses/by/4.0/).

applied sciences

MDPI

Article

Sound Radiation of Aerodynamically Excited Flat Plates into Cavities

Johannes Osterziel [1,*], Florian J. Zenger [2] and Stefan Becker [2]

[1] Department of Electrical, Electronic and Communication Engineering, Chair of Sensor Technology, Friedrich-Alexander University Erlangen-Nürnberg, Paul-Gordan-Straße 3/5, Erlangen 91052, Germany

[2] Department of Chemical and Biological Engineering, Institute of Process Machinery and Systems Engineering, Friedrich-Alexander University Erlangen-Nürnberg, Cauerstraße 4, Erlangen 91058, Germany; ze@ipat.fau.de (F.J.Z.); sb@ipat.fau.de (S.B.)

* Correspondence: johannes.osterziel@fau.de; Tel.: +49-9131-85-23359

Received: 22 August 2017; Accepted: 7 October 2017; Published: 14 October 2017

Abstract: Flow-induced vibrations and the sound radiation of flexible plate structures of different thickness mounted in a rigid plate are experimentally investigated. Therefore, flow properties and turbulent boundary layer parameters are determined through measurements with a hot-wire anemometer in an aeroacoustic wind tunnel. Furthermore, the excitation of the vibrating plate is examined by laser scanning vibrometry. To describe the sound radiation and the sound transmission of the flexible aluminium plates into cavities, a cuboid-shaped room with adjustable volume and 34 flush-mounted microphones is installed at the non flow-excited side of the aluminium plates. Results showed that the sound field inside the cavity is on the one hand dependent on the flow parameters and the plate thickness and on the other hand on the cavity volume which indirectly influences the level and the distribution of the sound pressure behind the flexible plate through different excited modes.

Keywords: aeroacoustics; flow-excited; vibration; sound radiation; acoustic coupling; turbulent; boundary layer

1. Introduction

There are noise sources everywhere in our environment, and these sounds are transferred in different ways to the human ear. The simplified generic model subject to this investigation reflects the type of sound excitation and transmission that occurs in numerous applications, e.g. in passenger vehicles, such as cars, buses, trains or planes. This undesirable sound is caused by excited flat structure elements. Today, people are constantly exposed to continuous noise which constitutes one of the most serious problems of our time [1]. Therefore, a growing need for personal acoustic shielding arises. The increasing travel time for great distances is indirectly leading to higher driving velocities to save time in turn. Thus, the flow noise develops to the dominating driving noise. In this context, there are numerous scientific studies discussing individual issues of the noise origination without considering the entire effect chain, starting with excitation, over the sound radiation und transmission to the sound detection.

Recent studies analyzed the flow-induced vibration of a plate with a crosswise prefixed obstacle plate and an additional external exciter with a varying frequency and amplitude [2]. External forces often have an effect on flow-excited plates. It turned out that both excitations are influencing the dynamics of the system. The frequencies of both test modes appear in the frequency spectrum at a low amplitude of the external harmonic excitation, whereas the external excitation in a coupled system is prevailing at a higher amplitude.

Concerning the sound radiation into a room, there were vibroacoustical analytical investigations on a model of a railway vehicle body [3]. The structure-induced noise radiation of flexible floor panels was analyzed and the natural frequencies and the sound pressure field inside the cavity were analytically determined. The result of the study is that significant resonance peaks arise from two concentrated forces on the plate if the dominant excitation frequency approaches the natural frequencies of the system. Moreover, the rise of the flow velocity led to an increase in the sound pressure inside the cavity. Regarding the spatial distribution of the sound pressure, it was found that the sound pressure above the excitation points takes remarkably higher values.

Numerous cases regarding the flow over different obstacles have already been investigated. Low importance has been attached to the sound radiation of a plate and the coupled distribution inside a cavity yet. However, two novel studies considering this subject were published this year. SHI et al. [4] focussed on the analytical modeling of a three-dimensional coupled acoustic system consisting of a cavity with a coupled flexible plate and a semi-infinite expanded field. The excitation of the system, including acoustic and structural-acoustic coupling, is generated inside the cavity by a point source. A solution method was found to predict the dynamic behavior of the entire system using Fourier series to express the sound pressure and the displacement of the plate. The results are validated through numerical simulations. This approach examined the issue which will be covered in detail in this paper, but only in an analytical way with an excitation in reverse direction without the influence of an external flow.

GANJI and DOWELL [5] investigated the sound transmission into and radiation from a cavity through a flexible plate structure in a noisy/thermal environment in a supersonic flow. Using nonlinear models for the pre- and the postflutter regions, equations for the flexible panel, cavity acoustics and turbulent boundary layer are developed. Based on several approaches and theories, the resulting equations are solved numerically. The results for the postflutter domain are high sound pressure levels in the lower-frequency regions and a Helmholtz resonator condition with homogenous sound pressure distribution inside the cavity.

While there are numerical and analytical studies, there are hardly any experimental investigations. This paper provides correlations between the excitation, the sound radiation and the sound detection. In particular, the flow is characterized, the plate vibration analyzed and the measured sound inside the cavity is described.

The aim of this study is to understand the effects of different influences on the sound radiation and the sound transmission of flexible plate structures with different measuring configurations. The effect chain of a clearly defined excitation should be understood in small scale and hereinafter transferred to special application fields. Moreover, an additional flexible plate could be added and different damping and absorbing materials could be analyzed to reduce the sound radiation and transmission.

2. Theory

Before analyzing the sound radiation of the plate, a determination of the boundary layer parameters is required to achieve a starting position with well-defined initial and boundary conditions. For illustrative purposes, a scheme of the theoretical setup is depicted in Figure 1.

To specify whether the boundary layer is laminar or turbulent, an analytical calculation of the dimensionless velocity profiles is required. Therefore, different approximate solutions for the wall shear stress τ_w are examined in detail. The first one has resulted from empirical measurements of a pipe flow and can be transferred to smooth plate walls due to similarity. The analogous approach for a turbulent boundary layer [6,7]

$$\frac{u_\infty}{u_\tau} = 8.75 \left(\frac{u_\tau \, \delta}{\nu} \right)^{\frac{1}{7}} \tag{1}$$

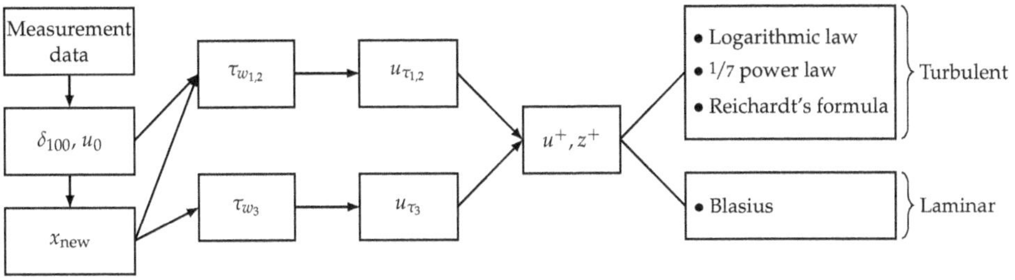

Figure 1. Scheme of the theoretical setup.

is solved for the friction velocity with the main flow velocity u_∞, the wall shear stress velocity u_τ, the boundary layer thickness δ and the kinematic viscosity ν. Using the following equation $\tau_w = \rho u_\tau^2$ with the density ρ yields:

$$\tau_{w_1} = 0.0225 \cdot \rho\, u_\infty^2 \left(\frac{\nu}{u_\infty \cdot \delta} \right)^{\frac{1}{4}} \tag{2}$$

The boundary layer thickness δ of the turbulent flow as a function of the running length x is developed through substitution in conjunction with the 1/7 power law, rearrangement and integration from 0 to x [6,8]:

$$\delta(x) = 0.371 \cdot x \cdot {Re_x}^{-\frac{1}{5}}. \tag{3}$$

with the Reynolds number based on the running length in flow direction Re_x. For Reynolds numbers ranging from 5×10^5 to 10^7, the local friction coefficient for a turbulent flow over an isothermal plate is correlated by [9,10]:

$$C_{f,x} = 0.0592 \cdot {Re_x}^{-\frac{1}{5}} \tag{4}$$

The friction coefficient as the dimensionless shear stress is defined as [9]:

$$C_{f,x} = \frac{\tau_w}{\frac{\rho}{2} \cdot {u_\infty}^2} \tag{5}$$

Equalizing the two Equations (4) and (5) results in the following wall shear stress τ_{w_2} as second equation:

$$\tau_{w_2} = 0.0296 \cdot \rho\, u_\infty^2\, {Re_x}^{-\frac{1}{5}} \tag{6}$$

We tried to approach the velocity distribution in the boundary layer to the real course of the experimental measurement data. Indirect turbulence models are used as a basis of the direct velocity modeling:

- **1/7 Power Law** [6,9]

$$u(z) = u_\infty \cdot \left(\frac{z}{\delta(x)} \right)^{\frac{1}{7}} \tag{7}$$

with the distance perpendicular to the plate z. The dimensionless velocity is calculated using the following equation $u^+ = \frac{u(z)}{u_\tau}$ with $u_\tau = \sqrt{\frac{\tau_w}{\rho}}$.

- **Logarithmic Law after Coles and Adapted by Golliard** [11,12]

The dimensionless velocity which follows the outer-layer corrected logarithmic law after Coles is given by:

$$u^+ = \frac{u(z-z_0)}{u_\tau} = \frac{1}{\kappa} \ln \left(Re_\delta \frac{u_\tau}{u_\infty} \xi \right) + B + \frac{\Pi}{\kappa} \omega(\xi) \quad (0 \le (z - z_0) \le \delta) \tag{8}$$

This indirect turbulence model uses a Reynolds number which is based on the boundary layer thickness δ:

$$Re_\delta = \frac{u_\infty \delta}{\nu} \tag{9}$$

The dimensionless wall distance ξ is defined as

$$\xi = \frac{z - z_0}{\delta}, \tag{10}$$

with the position of the wall z_0. The von Kármán constant is $\kappa = 0.4$ [11] and the boundary layer constant is $B = 5.1$. HINZE [13] fitted the data analytically from an empirical table after Coles and derived the following relation for the wake-function:

$$\omega(\xi) = 1 - \cos(\pi\xi) = 2\sin^2\left(\frac{\pi}{2}\frac{z}{\delta}\right) \tag{11}$$

The profile parameter Π is dependent on the pressure gradient and results in Equation (8) for $\xi = 1$. At this point the velocity is u_∞ and the z-coordinate corresponds to the boundary layer thickness δ.

$$\Pi = \frac{\kappa}{2}\frac{u_\infty}{u_\tau} - \frac{1}{2}\ln\left(Re_\delta \frac{u_\tau}{u_\infty}\right) - \frac{B\kappa}{2} \tag{12}$$

- **Reichardt's Formula** [14]

$$u^+ = \frac{1}{\kappa}\ln(1 + \kappa z^+) + 7.8\left[1 - \exp^{-\frac{z^+}{11}} - \frac{z^+}{11}\exp^{-0.33z^+}\right] \tag{13}$$

with the dimensionless wall distance $z^+ = \frac{z \cdot u_\tau}{\nu}$.

Regarding the laminar boundary layer, there is a solution in general form of the simplified approximated form of the boundary layer equations after Prandtl for the special case of a flat plate [9]. The appropriate derivation through a similarity clause was realized by Blasius [9]. In this case, another formula for the wall shear stress according to the laminar boundary layer is used:

$$\tau_{w_3} = 0.332 \cdot \rho\, u_\infty^2\, Re_x^{-\frac{1}{2}} \tag{14}$$

The running length x is defined so that it takes the value 0 at the front edge of the rigid plate, which is connected to the wind tunnel, and represents the x-position of the measured velocity profiles. This position is either directly or indirectly (by using the Reynolds number) required to calculate the wall shear stress. Moreover, this assumption states a boundary layer which is turbulent from the beginning and 'normally' developing in x-direction. However, the transition from laminar to turbulent in this investigation is artificially created by a roughness trip located at the inner wall of the nozzle. Thus, the transition occurs in a certain distance, which is displaced in the downstream direction and not unambiguously determinable. Consequently, a new x-position is reverse calculated by rearranging Equation (3):

$$x_{\text{new}} = \left(\frac{\delta_{100} \cdot u_\infty^{\frac{1}{5}}}{0.371 \cdot \nu^{\frac{1}{5}}}\right)^{\frac{5}{4}} \tag{15}$$

The boundary layer thickness δ_{100} and the velocity u_0 at the boundary layer border are directly estimated from the hot-wire measurements because usual definitions are not sufficient. For this case, it can not assumed that the boundary layer thickness is the position where u_0 takes a constant value of $0.99 \cdot u_\infty$. Therefore, the measurement data are fitted with a spline curve. To determine its gradient, a moving average with a window width of 50 samples is used due to the noise caused fluctuations of the flow velocity at the boundary layer border. At the beginning the gradient is high but it decreases constantly to almost zero. Below a gradient of 0.05, only minor velocity fluctuations occur which leads to a good approximation of the wall distance z for the boundary layer thickness δ_{100}.

3. Materials and Methods

This section mentions all the measurement methods with its associated measuring setups and evaluation procedures.

3.1. Experimental Setup

The following subsections deal with the experimental setups that have been utilized to measure the fluid flow and the sound field.

3.1.1. Flow Field

The measurements were performed in the aeroacoustic wind tunnel of the Chair of Sensor Technology at the Friedrich-Alexander University Erlangen-Nürnberg. The measuring section is located in an anechoic room with the dimensions 9 m × 6 m × 3.6 m. The general test setup is installed at a massive base frame built from aluminium profiles to ensure the structural stiffness. To determine the pressure coefficient at the plate surface, 64 holes with the diameter of 0.3 mm are drilled on the flow-exposed surface of the rigid plate, whereas on the other side a threated hole is drilled and tapped to connect pressure hoses to 4 Pressure Systems 16TC/DTC pressure scanners from the company Pressure Systems (Melbourne, Australia). The pressure holes are arranged in a straight streamwise line with a constant spacing from one another. The second measurement setup is shown in Figure 2. The replaced continuous rigid plate is equipped with a square opening in which a manufactured aluminium frame is inserted. The different flexible plates with the thickness $t = 1, 3$ and 5 mm are each bonded to the frame by an adhesive to avoid possible tensions in contrast to bolting. After installation of the frame, the plane surface of the flexible plate is arranged at the same height as the solid plate. The position of the flexible plate in relation to the nozzle outlet is set so that the entire surface of the plate structure is located in the core area of the free-stream at any time. This substructure ensures that the only way that sound is transferred into the cavity is through the flexible plate.

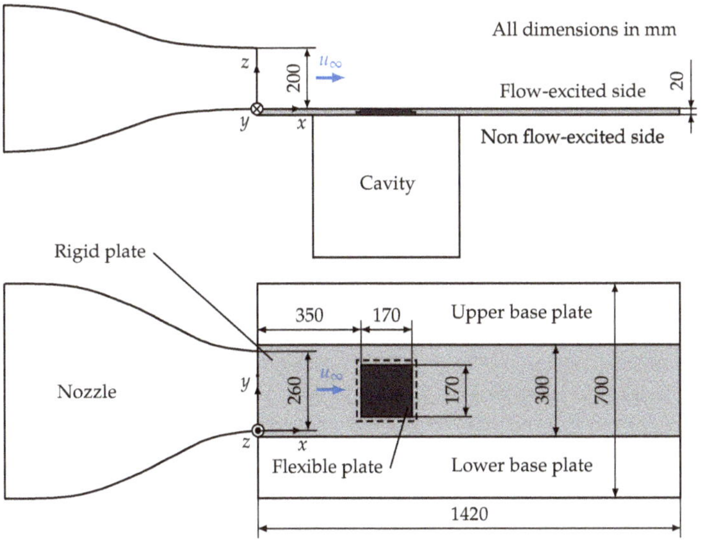

Figure 2. Schematic sketch of the experimental setup.

3.1.2. Sound Field

To investigate the sound radiation and transmission of the flexible plate more precisely, a cuboid-shaped room is installed at the back surface of the rigid plate. The outer walls of this cavity are built from 21 mm thick multiplex panels and equipped with microphones to analyze the radiated sound field. The complete mounting and the exploded view are depicted in Figure 3. The cavity volume can be varied by putting the rear panel from above in appropriate recesses at the insides of the sidepanels. The internal dimensions of the cavity are 450 mm × 217 mm × (255 mm or 435 mm). Consequently, there are two different values for the side length of the cavity, orthogonal to the plate structure. Accordingly, the two obtained different minimum and maximum possible cavity volumes with a room depth of 255 mm and 435 mm are $V_{min} \approx 0.0249\,m^3$ and $V_{max} \approx 0.0425\,m^3$.

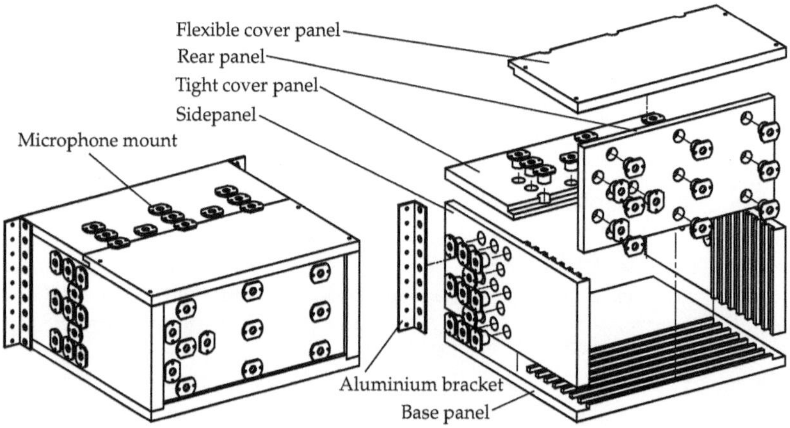

Figure 3. Assembly and exploded view of the cavity.

3.2. Measurement Technology and Methods

The wall pressure was measured over a period of 60 s with a sampling frequency of 100 Hz using the pressure system DTC Initium working with the DTC-technology (Digital Temperature Compensation). To determine the pressure coefficient, 64 positions on the plate in the direction of the flow are simultaneously recorded. The determination of the velocity and of the turbulence intensity is forming the basis for the flow characterization. To approach the true value of the flow velocity as accurately as possible, the velocity regulating differential pressure transducer Model 239 from Setra Systems (Boxborough, Middlesex, MA, USA) is calibrated with the Prandtl Probe TPL-03-200 from KIMO Instruments (Montpon Ménestérol, France) working with very high accuracy. On this basis, the velocity field and the turbulence intensity distribution perpendicular to the flow direction in a certain distance to the nozzle outlet are measured by this Prandtl probe and a hot-wire anemometer with the miniature wire probe 55A53 from Dantec Dynamics (Skovlunde, Denmark).

Using a traversing system, a zero point in the corner of the nozzle outlet is set as a reference for the 108 measurement points. They are arranged on a planar grid, parallel to yz-plane with equal spacing. The velocity profiles were measured at distances $x = 100, 350$ and 520 mm from the nozzle outlet with a sampling rate of 5 kHz and a measuring period of 20 s. The turbulence intensity is determined in the same planes.

For the examination of the boundary layer, the boundary layer probe 55P15 from Dantec Dynamics is necessary in terms of measurements in direct wall proximity. The corresponding measurement setup with the nozzle, the flow-excited plate and the traversing system is shown in Figure 4. The measurement distances to the nozzle are chosen between the position of the front edge ($x = 350$ mm) and the rear edge ($x = 520$ mm) of the installed flexible plate. More precisely, different not equidistant points in z-direction perpendicular to the plate are measured. While approaching towards the wall, the distance is successively decremented. The traversing system automatically moves the probe to all points till the contact point between probe and plate is reached.

Figure 4. Measurement setup and enlargement of the boundary layer probe.

The microphones in the side panel and the top panel are arranged in such a way that they are able to collect the generated noise at anytime independent of the rear panel position. Moreover, one microphone row is placed in the middle of the minimum and the maximum volume in each case. The cavity volume and the microphone measuring positions are varied for the same plate thickness of the flexible plate structure. This results in different measuring configurations with the following variation possibilities:

- Cavity volume: minimum and maximum
- Microphone arrangement: left, top and rear
- Flow velocity: $45\,\mathrm{m\,s^{-1}}$
- Plate thickness: 1 mm, 3 mm and 5 mm

Additionally, a frame, milled from solid with a thickness of 20 mm, is precisely closing the square opening in the rigid aluminium plate and has been prepared for comparative purposes. It can be assumed that there is no noise transmitted through this frame.

To detect structural vibrations of the flexible plate and to establish a correlation with the sound radiation in the cavity, Laser Scanning Vibrometer measurements are performed. The method is based on the Doppler-effect where two coherent light beams with different path lengths overlap and result in optical interference. The velocity of a certain point on the vibrating plate is determined through the known wavelength of the laser beam and the frequency shift due to the motion of the surface (Figure 5).

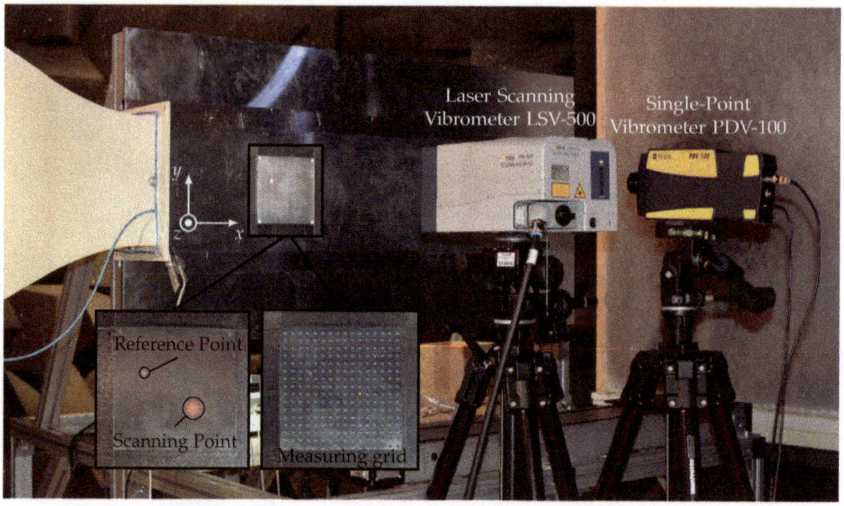

Figure 5. Measurement setup with Laser Scanning Vibrometers.

A square-shaped grid with 256 measuring points is generated and placed on the surface of the flexible plate using the Laser Scanning Vibrometer PSV-500 from Polytec (Waldbronn, Germany) with an automatic scanning mirrors system. To accurately replicate the structural modes of the plate, the phase relation of each measuring point to one fixed point is required. Therefore, the Single-Point Vibrometer PDV-100 from Polytec is connected to the first reference input channel of the data acquisition card of the PSV-500 Scanning System. Flush-mounted microphones are also used as reference signals.

4. Experimental Results

The experimental results of all measurements are presented and interpreted in this section.

4.1. Flow Characterization

Initially, the flow directly influencing and exciting the flexible aluminium plate was examined in detail in the following two subsections. In particular, different flow parameters and the boundary layer on the flow-excited surface of the flexible plate have been investigated.

4.1.1. Flow Parameters

In general, the wind pressure merely acts towards the plate surface and thus, the pressure coefficient achieves positive values. The results show that first there is a sharp decline till the pressure coefficient is reaching a certain level. This suggests that the flow is not fully turbulent developed yet, but the transition area between laminar and turbulent flow is still present. A rough calculation of the critical Reynolds number confirms this assumption. In conclusion, there is no boundary layer with zero pressure gradient because the pressure coefficient is not close enough to zero. However, the values are only slightly increasing above 0.01 so that the deviation is relatively small. The relevant range between the front edge and the trailing edge of the flexible plate shows the least fluctuating pressure coefficient with values between 0.005 and 0.01. Consequently, the boundary layer along a flat plate is characterized by the fact that the velocity profiles measured at different distances to the leading edge of the rigid plate are similar.

The spatial distribution and the accuracy of the flow velocity is very important in the aeroacoustic wind tunnel because the validity of the measurements is strongly dependent on whether a uniform velocity field with nearly constant velocity in the core region is formed in the test section. The resulting velocity profiles are depicted in Figure 6 by means of contour plots linked to a simplified CAD model of the nozzle and the rigid plate. It is shown that the core area of the flow with a nearly constant velocity is shrinking, whereas the mixing zone in the transition area to the resting medium is broadening. The flow-excited flexible plate, which is exactly located between the two velocity profiles at x = 350 and 520 mm, is clearly lying in the core flow.

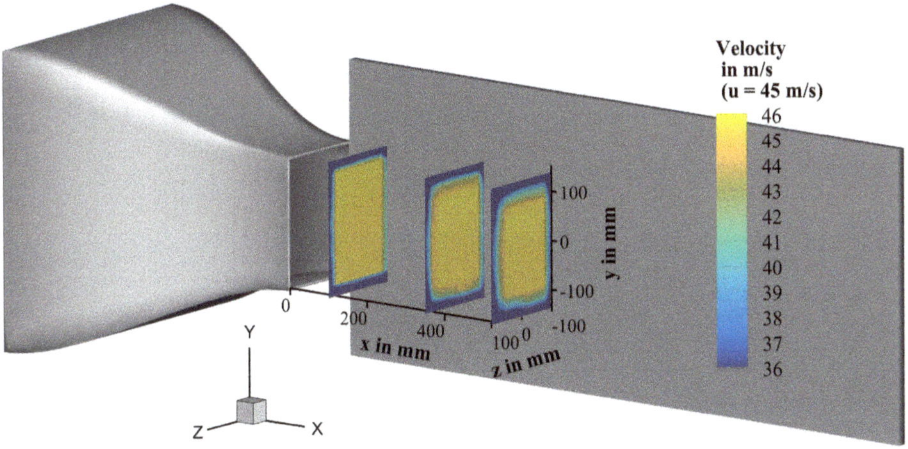

Figure 6. Velocity profile at 45 m s^{-1} with distances x = 100, 350 and 520 mm from the nozzle outlet.

The turbulence intensity for isotropic turbulence with equal average velocity fluctuations in all three dimensions $u_x'^2 = u_y'^2 = u_z'^2$ [9,15] is

$$T_u = \frac{\sqrt{u_x'^2}}{u_\infty} \times 100 \quad \text{in}\,\%, \tag{16}$$

The distribution of the increasing turbulence intensity from 0 % towards the edges is similiar to the decreasing velocity from 45 m s^{-1} (Figure 6). Moreover, the turbulence intensity is identically depicted as the velocity profiles (Figure 7) at the same position. The turbulence intensity values at the edges increase with distance to the nozzle. Accordingly, a contour with relatively small width and increased turbulence intensity, which is recreating the form of the nozzle outlet, is growing in width towards the middle of the profile. The first contour plot shows a high turbulence intensity at the edges. Considering the contour plots at the position x = 350 mm and 520 mm, the turbulence intensity at the edges decreases with increasing distance to the nozzle. As already mentioned in connection with the flow velocity, a good comparability and transferability of the measurement results can be provided because the area of the flexible plate structure is exposed to low turbulence.

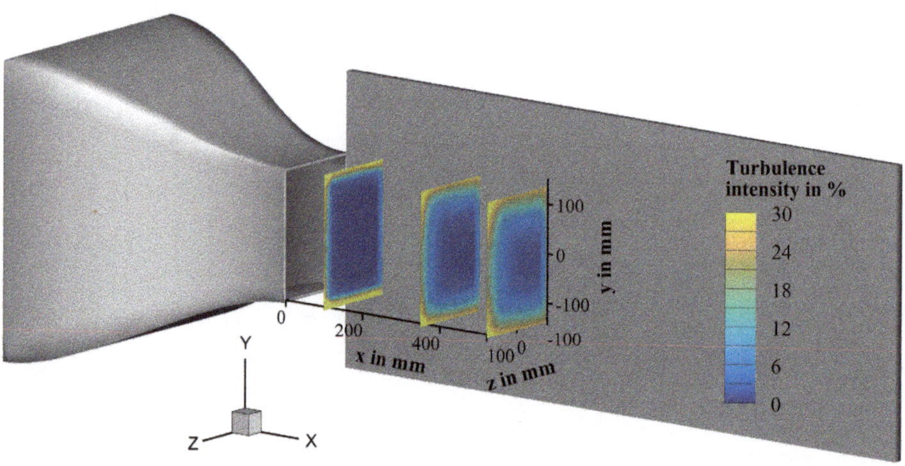

Figure 7. Turbulence intensity at 45 m s^{-1} with a distance x = 100, 350 and 520 mm from the nozzle outlet.

4.1.2. Boundary Layer

The dimensionless velocity profiles of the measurement, of the three indirect turbulence models and of the approach after Blasius (refer to Section 2) for a distance to the nozzle outlet of $x = 350$ mm are shown in Figure 8. The turbulence models correlate in exponential curve characteristics compared with the measurement in contrast to the approach after Blasius which shows a linearly increasing course at the beginning changing to an exponential growth till the maximum value of u^+. This result suggests that the boundary layer is turbulent.

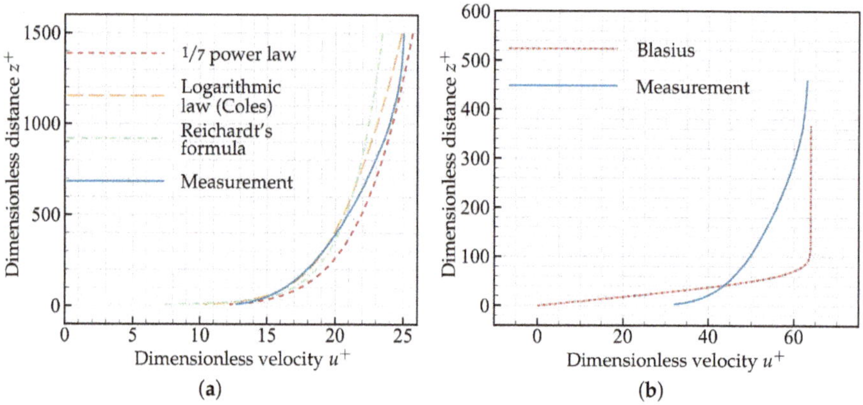

Figure 8. Dimensionless velocity profiles at 45 m s^{-1} with a distance of x = 350 mm from the nozzle outlet. (**a**) Turbulent presumed boundary layer in case of shear stress 1; (**b**) Laminar presumed boundary layer.

To evaluate the quality of the indirect turbulent models more clearly, another depiction of the dimensionless velocity profiles of the turbulent presumed boundary layer is presented in Figure 9. Besides this, the graphs of the logarithmic overlap law and of the laminar sublayer are added. The 1/7 power law reproduces the curve shape of the measurement most closely at the beginning, whereas the logarithmic law after Coles overlaps the measurement between a dimensionless distance from $z^+ = 90$ to 200. Towards the end, a plateau with constant dimensionless velocity is to be noticed which means that this section is already outside of the boundary layer. The curve progressions of the logarithmic law and the logarithmic overlap law are the same till reaching the outer part of the boundary layer. There, the law of the wake as the third summand added to the logarithmic overlap law takes effect and describes the outer part in the corrected equation of Coles (Equation (8)). The laminar sublayer completely deviates from the other curves.

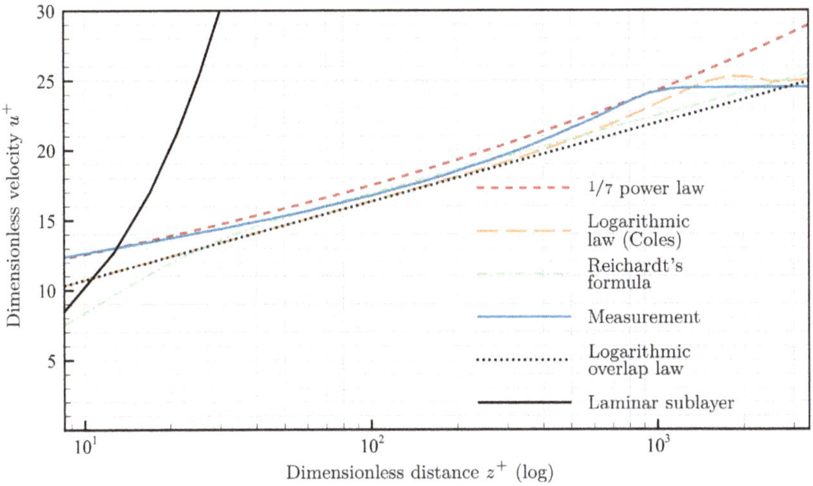

Figure 9. Dimensionless velocity profiles of the turbulent presumed boundary layer at 45 m s^{-1} in case of shear stress 1 with a distance of x = 350 mm from the nozzle outlet.

The shape factor H_{12} as the ratio of the displacement thickness δ_1 to the momentum loss thickness δ_2 is used to characterize the boundary layer developing on a flat plate. The integral calculation of the displacement thickness δ_1 and the momentum loss thickness δ_2 is done by a spline interpolation with a following exact numerical integration using global adaptive quadrature and preset fault tolerance [16]. The values of H_{12} converge towards 1.3 despite varying values for δ_1 and δ_2 for different x-positions. For laminar flows on a flat plate, H_{12} is typically about 2.59. At the transition point it falls to a considerably smaller number range between 1.4 and 1.0 [9,17]. CLAUSER [18] concluded that H_{12} is dependent on the Reynolds number Re_x based on the running length x and on the skin friction coefficient C_f for equilibrium boundary layers. In this case of constant pressure profiles, for a range of $10^6 < Re_x < 10^7$, the values of H_{12} are expected to be between 1.27 and 1.35 with C_f between 0.025 and 0.036. According to HEAD, COLES and Smith and Walker [19] H_{12} assumes higher values between 1.3 and 1.42 in the same range of Re_x. NIKURADSE [20] concluded after measuring the boundary layer for a pipe flow that H_{12} = 1.3 is constant in the measurement range of $1.7 \times 10^6 \leq Re_x \leq 1.8 \times 10^7$. All boundary layer parameters are shown in Table 1. The resulted values of H_{12} clearly prove that there is a turbulent flow in the entire area of the flexible flow-excited plate. Moreover, the calculated x-distance x_{new} from Equation (15) is nearly twice as high as the real x-position. The boundary layer thickness δ_{100} increases with the distance to the nozzle.

Table 1. Boundary layer parameters for the analytical approaches.

x-pos.	x_{new}	u_∞	u_{τ_1}	δ_{100}	δ_1	δ_2	H_{12}	$Re_x =$	$Re_{\tau_1} =$	$Re_{\delta_2} =$
mm	mm	m s^{-1}	m s^{-1}	mm	-	-	-	$u_\infty x_{new}/\nu$	$u_\tau \delta_{100}/\nu$	$u_\infty \delta_2/\nu$
350.0	702.2	45.00	1.80	14.44	1.80	1.38	1.30	1.91×10^6	1571	3.76×10^6
392.5	721.7	45.01	1.79	14.77	1.89	1.45	1.31	1.96×10^6	1598	3.93×10^6
435.0	807.7	45.01	1.78	16.18	2.07	1.58	1.31	2.18×10^6	1722	4.26×10^6
477.5	859.4	45.01	1.77	17.01	2.22	1.70	1.31	2.31×10^6	1796	4.58×10^6
520.0	977.8	45.01	1.74	18.87	2.41	1.83	1.31	2.63×10^6	1963	4.92×10^6

The measurement data is fitted using the approach from Golliard [12] to reproduce the curve progression of the dimensionless velocity most accurately. The generated curve points based on the spline interpolation are entered as input data for the curve fitting using the least squares method. The logarithmic law from Equation (8) is the objective function, in which u_∞ is replaced by the velocity at the boundary layer border from the hot-wire measurements u_0. The z-distance is the passed dependent variable and the four fit variables are u_0, $u_{\tau_{Go}}$, δ_{100} and z_0 with the Levenberg-Marquardt algorithm as numeric optimization method [21,22]. Another curve fitting with constant u_0 and δ_{100} and only $u_{\tau_{Go_2}}$ and z_0 as fit parameters is performed to verify the quality of the results.

Comparing the wall shear stress resulting from the analytical calculation (u_{τ_1}, u_{τ_2}) with those from the numerical approximation ($u_{\tau_{Go}}$, $u_{\tau_{Go_2}}$), one can recognize different values as shown in Table 2. However, relating to their respective velocities (u_∞, u_0), nearly equal values are obtained. It can thus be concluded that the Equations (2) and (6) for the wall shear stresses τ_{w_1} and τ_{w_2} provide a very good approximation. The adapted running length x_{new} (Equation (15)) and the boundary layer thickness δ_{100} (Equation (3)) are used for the calculation. Without measurement data it is impossible to calculate or predict the boundary layer parameters in case of an artificially induced turbulent boundary layer, whereas the adjusted analytical approaches solely suffice to characterize the boundary layer with the existance of measurement data.

Table 2. Comparison of the boundary layer parameters to analytical approaches and to the curve fitting beyond the approach of Golliard.

x-pos.	u_0	u_{τ_1}	u_{τ_2}	$u_{\tau_{Go}}$	$u_{\tau_{Go_2}}$	$\frac{u_{\tau_1}}{u_\infty}$	$\frac{u_{\tau_2}}{u_\infty}$	$\frac{u_{\tau_{Go}}}{u_0}$	$\frac{u_{\tau_{Go_2}}}{u_0}$
mm	m s^{-1}	m s^{-1}	m s^{-1}	m s^{-1}	m s^{-1}	-	-	-	-
350.0	44.11	1.799	1.823	1.758	1.792	0.040	0.041	0.040	0.041
392.5	43.87	1.795	1.819	1.733	1.763	0.040	0.040	0.039	0.040
435.0	44.44	1.776	1.799	1.730	1.767	0.039	0.040	0.039	0.040
477.5	43.99	1.765	1.789	1.704	1.739	0.039	0.040	0.039	0.040
520.0	43.78	1.743	1.766	1.664	1.707	0.039	0.039	0.038	0.039

4.2. Flow-Induced Sound Radiation into the Cavity

To determine the influence of the single measuring configurations and to compare and assess those effects, there are only one or two varying parameters. All figures are related to microphone position number 5 which is each located in the middle of the rear, side and top panel of the cavity (Figure 10). Unless otherwise specified, the maximum volume and the microphone position on the rear panel is chosen in the following. The lower cut-off frequency of the cavity as an acoustic chamber with maximum volume is 2866 Hz. Above this frequency, a strong superpositioning of natural frequencies occur which is apparent in the following sound pressure level depictions.

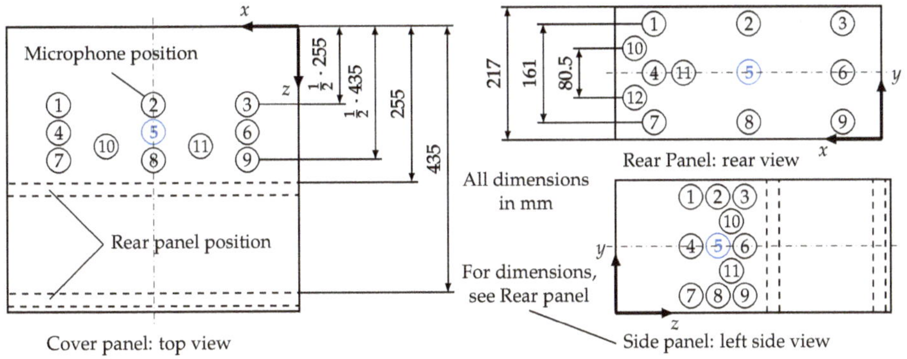

Figure 10. Sketch of the microphone positions on the cavity walls.

Figure 11 shows the effects of different microphone positions on the sound pressure levels. The measuring planes (Figure 11a), the distance from the flexible plate (Figure 11b) and finally the microphone position in the rear panel (Figure 11c) are varied. The frequency range is from 100 to 1000 Hz because that is the interesting area where significant modifications arise. The sound pressure level spectrum is plotted against the frequency for the three microphone positions on the different cavity walls.

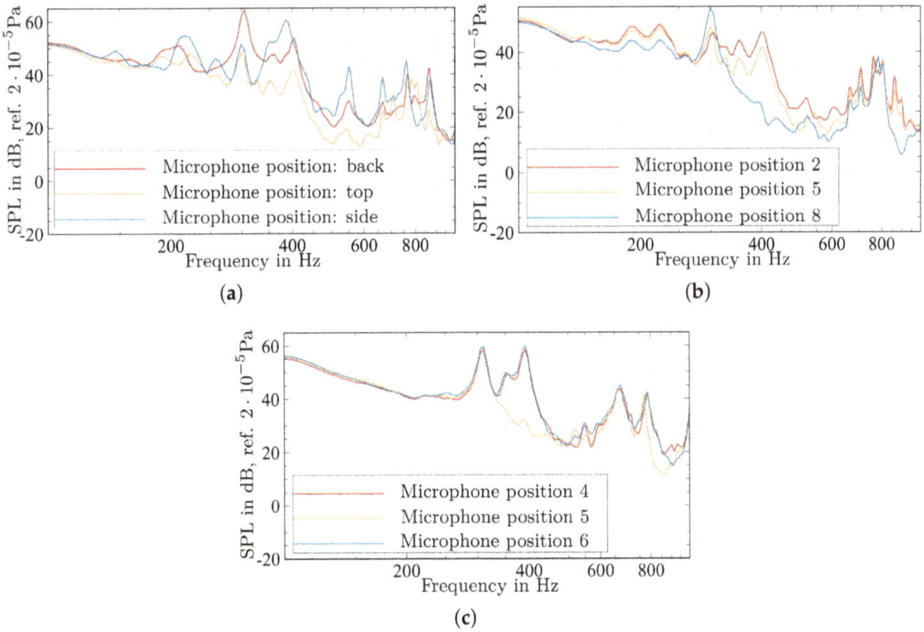

Figure 11. Sound Pressure Level at variation of the microphone position (plate thickness $t = 1$ mm). (**a**) Different measuring surfaces; (**b**) On the cover panel in z-direction; (**c**) On the rear panel in x-direction (minimum volume).

Common peaks appear in Figure 11a at 310 Hz, in the range of 400 Hz, 550 Hz and at higher frequencies, too. Below 300 Hz, the curve characteristics differ considerably because the frequency is below the frequency of the first room mode of the cavity and thus not influenced by the room.

The second figure (Figure 11b) depicts the different distances from the flexible plate with position 2 as the nearest position to the plate. Between 300 and 500 Hz, the sound pressure level measured at position 8, located at half of the cavity length, is significantly lower than those of the other positions. This is because special acoustic waves, whose wavelength correspond to the half room length, form nodes and thus sound pressure minima. A greater distance to the flexible plate leads to a lower sound pressure level at low frequencies with the exception of a few peaks.

Finally, different x-positions, describing the distances from the nozzle outlet in streamwise direction, are selected and located on the rear panel. The distance increases with a decreasing microphone number while number 5 exactly represents the perpendicularly displaced center of the flexible plate. There is a lower sound pressure level between the microphone positions 4 and 6 compared with 5 between 330 and 450 Hz and between 770 and 900 Hz. The responsible axial modes run in negative z-direction from the flexible plate and produce sound pressure minima at position 5. The curve shapes up to a frequency of 330 Hz in the low-frequency range are similar to each other.

Changing the rear panel position and thus the cavity volume implicates different developing sound pressure levels (Figure 12). The maximum at 210 Hz only occurs with maximum volume although cavity modes do not propagate below the first axial mode at approximately 400 Hz. This sound superelevation is dependent on the room size and position. The small peak at a frequency of 387 Hz can be explained by the second axial room mode for the minimum cavity volume.

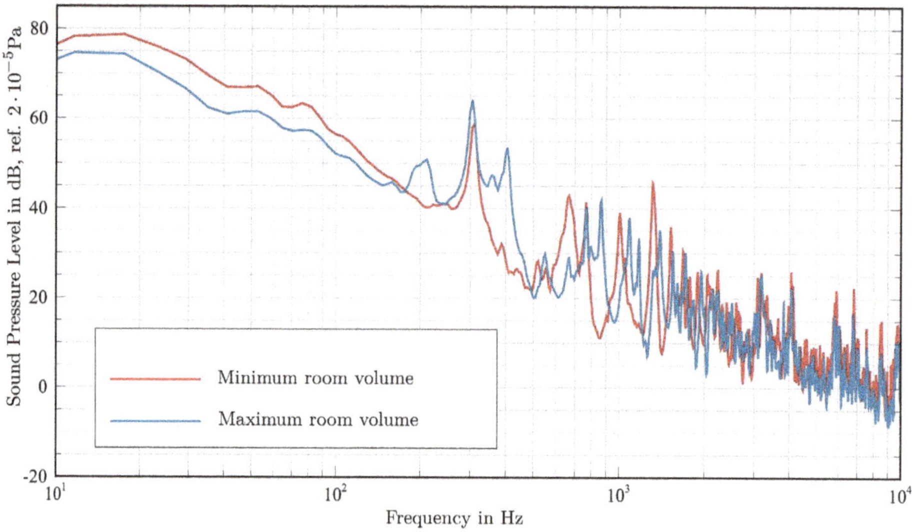

Figure 12. Sound Pressure Level at variation of the room volume.

The next step was to study the influence of plate thickness variation which is depicted in Figure 13. There is a considerable difference in the development of the sound pressure level at low frequencies for a plate thickness of 1 mm and the remaining thicknesses. The shape modes of the thinnest plate are excited and radiate sound into the cavity. Furthermore, the sound transmission through this plate is the highest among the other thicker plates. The first two peaks at 160 and 210 Hz appear regardless of the thickness and only for maximum volume. Subsequently, their development is not associated with the natural frequency of the thinnest plate. An immense sound pressure level variation is shown at 310 Hz.

Whereas there are only small impacts on the medium thick plates at this point, a huge positive peak is shown for the 1 mm thick plate and on the contrary a negative peak for the 20 mm thick solid plate. The short-circuit area and the piston-diaphragm area are located below the first natural frequency of the thinnest plate at 217.4 Hz. Thus, the increase can be explained by the sound transmission through the plate structure. At higher frequencies above 3000 Hz, the sound pressure level is increasing with a reduction of the plate thickness due to a different degree of oscillation excitation and varying sound transmission strength. Additionally, the illustrated frequency spectrum is extended to 20 kHz to show the sound pressure increase from approximately 12,800 Hz for the 1 mm thick plate. This is caused by the coincidence frequency above which a full sound radiation occurs. This effect is superimposed by other sound sources for the remaining plate thicknesses. Finally, the common peak at 550 Hz is due to the first tangential room mode.

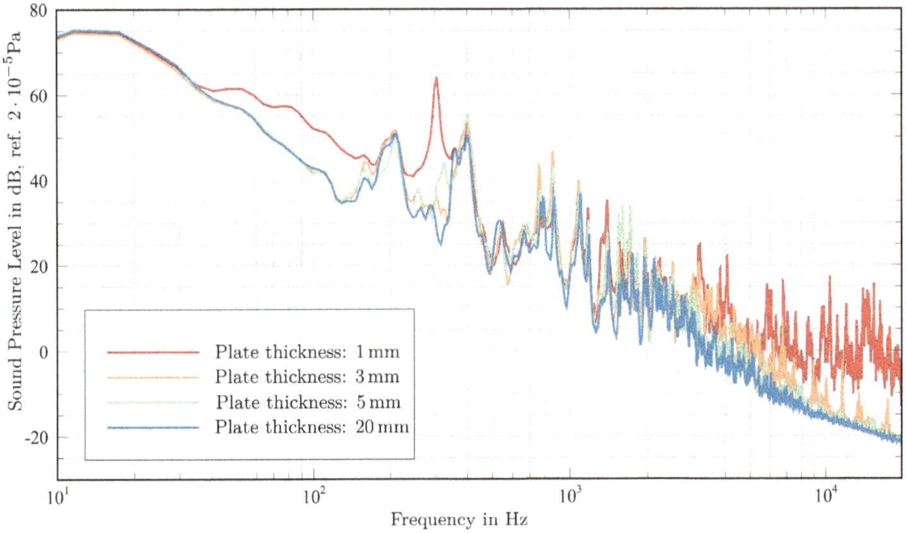

Figure 13. Sound Pressure Level at variation of the plate thickness.

To evaluate the mutual influence of the vibrating plate and the cavity, the auto power spectral density (APSD) of the surface normal velocity of the single-point vibrometer and the sound pressure level (SPL) of the microphone which is located in the middle of the rear panel are depicted together in Figure 14. The APSD is calculated as follows:

$$\mathrm{APSD} = 10\,\log_{10} \frac{\frac{1}{(\text{signal length}) \cdot (\text{sampling frequency})} \cdot |\mathrm{fft}(x)|^2}{10^{-18}} \tag{17}$$

with the signal length of 3200 and the sampling frequency of 2000 Hz. They are both converted to the unit dB but have two different reference values concerning the level calculation. There are 9 vertical dashed red lines representing the highest peaks of the APSD and 4 vertical dash dotted blue lines which show peaks of the SPL and small influences in the APSD at identical frequencies.

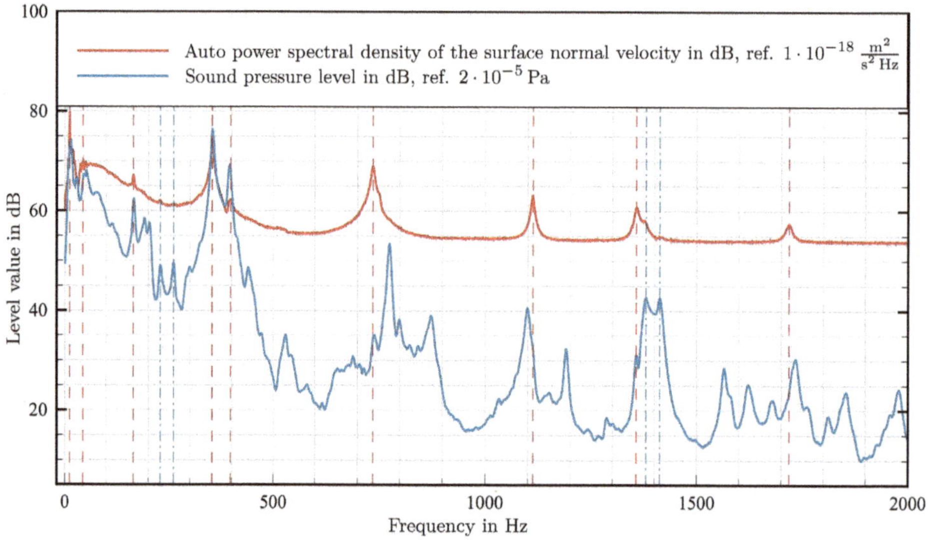

Figure 14. Comparison between structural vibration and sound radiation.

In the low frequency range till approximately 400 Hz, numerous maximum values of both curves appear at the same frequency in contrast to higher frequencies. Plate vibrations which occur below the first natural frequency of the plate at 353.125 Hz develop due to coupling with the entire frame structure of aluminium and pressurisation which is caused by low-frequency vortices.

The first peak at $f = 11.875$ Hz which is depicted in Figure 15a results from a vibration of the entire system which can be derived from the chronological sequence of the plate vibration. Every measurement point of the plate is oscillating with the same amplitude and phase. In the second Figure (Figure 15b), a superimposition between the vibration of the plate and a the formation of a kind of structural oscillation occurs although the first shape mode firstly appears at a higher frequency. The front area of the flow-excited plate, which is closer to the nozzle outlet, is considerably stronger oscillating upwards and downwards towards the surface normal than the rear region. The movement of the plate's rear edge is lagging behind the front edge like the phase-shifted points of a standing wave. The plate tilts whenever the front or the rear edge reaches the maximum oscillation amplitude. It is exposed to a compressive and tensile load caused by vortex structures with a long wavelength. The maximum value of the APSD is not clearly noticeable and has a higher peak width. The third peak of the APSD and the SPL results in a vibration form $f = 164.375$ Hz in Figure 15c. The bending of the plate moves in streamwise direction. One reason for the occuring mode shape across a wide frequency range which is similar to the first shape mode at $f = 353.125$ Hz (Figure 15d) is the increasing convection velocity with increasing frequency in the low frequency region [23]. Thus, the wavelength of the large eddies is slightly changing, whereby the effect on the plate remains similar. The vibration of the flexible plate leads to a sound radiation into the cavity which occurs at the first shape mode. Conversely, the first axial room mode at $f = 397$ Hz in Figure 15e appears in form of a peak in the SPL of the microphones and the APSD of the surface normal velocity. The plate is oscillating, induces the room mode, which in turn excites the plate to vibrate at the same frequency. The same phenomenon arises on a smaller scale at the frequencies which are highlighted by the 4 vertical dash dotted blue lines. The second shape mode at $f = 736.25$ Hz in Figure 15f has similarly to the following plate modes (Figure 15g–i) only a minor impact on the sound pressure level in the cavity.

In summary, the first shape mode has the greatest influence on the sound pressure level in the cavity. Moreover, the larger peaks of the APSD of the surface normal velocity match with those of the SPL in the low frequency region. The vibrations of the plate or the entire system with a higher amplitude are transferred into the cavity and influence the inner sound pressure level.

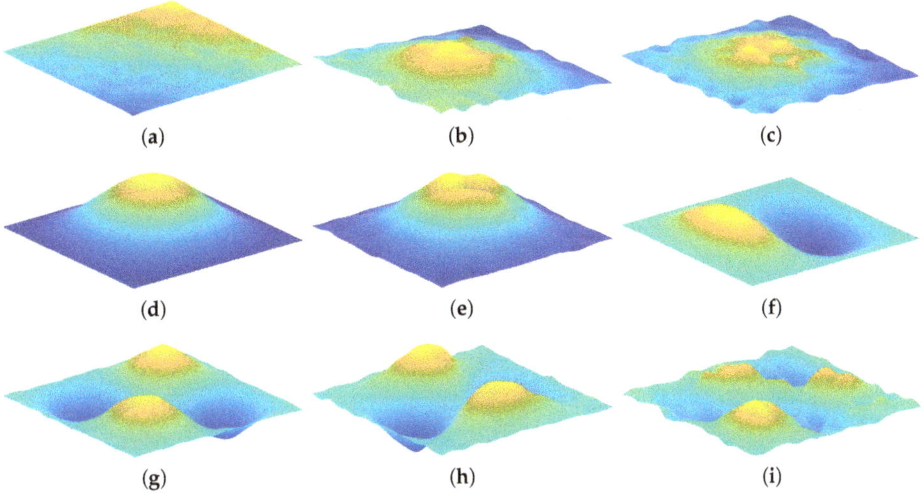

Figure 15. Plate vibration forms (plate thickness $t = 1\,\mathrm{mm}$). (**a**) Plate vibration at $f = 11.875\,\mathrm{Hz}$; (**b**) Plate vibration at $f = 45\,\mathrm{Hz}$; (**c**) Plate vibration at $f = 164.375\,\mathrm{Hz}$; (**d**) First shape mode at $f = 353.125\,\mathrm{Hz}$; (**e**) Plate vibration at $f = 396.875\,\mathrm{Hz}$; (**f**) Second shape mode at $f = 736.25\,\mathrm{Hz}$; (**g**) Third shape mode at $f = 1111.875\,\mathrm{Hz}$; (**h**) Forth shape mode at $f = 1356.875\,\mathrm{Hz}$; (**i**) Fifth shape mode at $f = 1718.125\,\mathrm{Hz}$

5. Conclusions

In this paper, measurements to characterize the flow and to determine the vibration and sound radiation of flexible plate structures into cavities are recorded. Moreover, the corresponding results are discussed. In summary, the flow is clearly defined by the analytical and numerical approaches. The velocity profiles and the turbulence intensity distribution reproduce the rectangular geometry of the nozzle outlet, are uniformly developed, but show wider edges with deviation from the optimum range. However, the flow-excited flexible plate structure is completely located in the core area of the open jet. The type of the flow along the plate is demonstrably a turbulent boundary layer which is proved by the shape factor H_{12}. To describe the boundary layer through the measurement data, the adapted analytical approaches suffice. The analysis of the flow-induced sound radiation into the cavity yields to different results depending on the measuring configuration. The highest sound pressure levels in the cavity are detected at the thinnest plate $t = 1\,\mathrm{mm}$. This can be seen both in the low frequency and the high frequency region and especially at 310 Hz. With the laser scanning vibrometer measurements, correlations between the vibration of a plate, the sound radiation and the retroactive effect of the cavity could be captured.

The aim of all the efforts is to understand the entire effect chain from the plate excitation to the sound detection for which fundamental knowledge is required. These includes basic experiments which were carried out in the present study. The measurements should serve as a benchmark for future tests. Adding another flexible plate behind the first one and testing different materials could provide more knowledge about multilayer constructions like already in use in nearly all vehicles.

Acknowledgments: This work was supported by Philipp Winter who contributed analysis tools.

Author Contributions: Johannes Osterziel and Florian J. Zenger conceived and designed the experiments; Stefan Becker contributed professional considerations with his sound knowledge. Johannes Osterziel performed the experiments and has been supported by Florian J. Zenger. Johannes Osterziel analyzed the data and wrote the paper. Florian J. Zenger and Stefan Becker were involved in the conception, development and correction of the written composition.

Conflicts of Interest: The authors declare no conflict of interest.

Abbreviations

The following abbreviations are used in this manuscript:

APSD	Auto Power Spectral Density
CAD	Computer Aided Design
CTA	Constant Temperature Anemometry
DTC	Digital Temperature Compensation
LSV	Laser Scanning Vibrometry
SPL	Sound Pressure Level

References

1. Schulte-Fortkamp, B.; Jakob, A.; Volz, R. *Lärm Im Alltag-Informationsbroschüre Zum Tag Gegen Lärm*; Deutsche Gesellschaft für Akustik e.V. (DEGA): Berlin, Germany, 2011.
2. Purohit, A.; Darpe, A.K.; Singh, S. Experimental Investigations on Flow Induced Vibration of an Externally Excited Flexible Plate. *J. Sound Vib.* **2016**, *371*, 237–251, doi:10.1016/j.jsv.2016.02.039.
3. Sadri, M.; Younesian, D. Vibro-Acoustic Analysis of a Coach Platform under Random Excitation. *Thin Walled Struct.* **2015**, *95*, 287–296, doi:10.1016/j.tws.2015.07.008.
4. Shi, S.; Su, Z.; Jin, G.; Liu, Z. Vibro-Acoustic Modeling and Analysis of a Coupled Acoustic System Comprising a Partially Opened Cavity Coupled with a Flexible Plate. *Mech. Syst. Signal Process.* **2018**, *98*, 324–343, doi:10.1016/j.ymssp.2017.04.045.
5. Ganji, H.F.; Dowell, E.H. Sound Transmission and Radiation from a Plate-Cavity System in Supersonic Flow. *AIAA J.* **2017**, 1–24, doi:10.2514/1.C034309.
6. Schade, H.; Kunz, E. *Strömungslehre*, 3rd ed.; Walter de Gruyter: Berlin, Germany, 2007; ISBN 978-3-11-018972-8.
7. Kaufmann, W. *Technische Hydro- und Aeromechanik*, 3rd ed.; Springer: Berlin/Heidelberg, Germany, 1963; ISBN 978-3-662-13102-2.
8. Som, S.K. *Introduction to Heat Transfer*; PHI Learning Pvt. Ltd.: New Dehli, India, 2008; ISBN 978-81-203-3060-3.
9. Schlichting, H.; Gersten, K. *Grenzschicht-Theorie*, 10th ed.; Springer: Berlin/Heidelberg, Germany, 2006; ISBN 978-3-540-23004-5.
10. Incropera, F. *Fundamentals of Heat and Mass Transfer*, 6th ed.; John Wiley & Sons: Hoboken, NJ, USA, 2007; ISBN 978-1-118-13727-7.
11. Coles, D. The Law of the Wake in the Turbulent Boundary Layer. *J. Fluid Mech.* **1956**, *1*, 191–226, doi:10.1017/S0022112056000135.
12. Golliard, J. *Noise of Helmholtz-Resonator like Cavities Excited by a Low Mach-Number Turbulent Flow*; Flow Acoustics Group, Acoustic and Vibrations Division, Institute of Applied Physics (TPD), Netherlands Organisation for Applied Scientific Research (TNO): The Hague, The Netherlands, 2002; ISBN 978-9-067-43964-0.
13. Hinze, J.O. *Turbulence*, 2nd ed.; McGraw-Hill: New York, NY, USA, 1975; ISBN 978-0-070-29037-2.
14. Reichardt, H. Vollständige Darstellung der turbulenten Geschwindigkeitsverteilung in glatten Leitungen. *J. Appl. Math. Mech.* **1951**, *31*, 208–219, doi:10.1002/zamm.19510310704.
15. Durst, F. *Grundlagen der Strömungsmechanik: Eine Einführung in die Theorie der Strömung von Fluiden*; Springer: Berlin/Heidelberg, Germany, 2006; ISBN 978-3-540-31324-3.
16. Shampine, L. Vectorized Adaptive Quadrature in MATLAB. *J. Comput. Appl. Math.* **2008**, *211*, 131–140, doi:10.1016/j.cam.2006.11.021.
17. Schimmelpfennig, S. Aeroakustik von Karosseriespalten. Ph.D. Thesis, Friedrich-Alexander Universität Erlangen-Nürnberg, Erlangen, Germany, 2015.

18. Clauser, F. Turbulent Boundary Layers in Adverse Pressure Gradients. *J. Aeronaut. Sci.* **1954**, *21*, 91–108, doi:10.2514/8.2938.
19. Head, M.R. *Entrainment in the Turbulent Boundary*; HM Stationery Office: London, UK, 1960.
20. Nikuradse, J. *Turbulente Reibungsschichten an der Platte*; ZWB, Oldenbourg: Munich/Berlin, Germany, 1942.
21. Levenberg, K. A Method for the Solution of Certain Non-Linear Problems in Least Squares. *Quart. Appl. Math.* **1944**, *2*, 164–168, doi:10.1090/qam/10666.
22. Marquardt, D. An Algorithm for Least-Squares Estimation of Nonlinear Parameters. *SIAM J. Appl. Math.* **1963**, *11*, 431–441, doi:10.1137/0111030.
23. Mccolgan, C.J.; Larson, R.S. *Mean Velocity, Turbulence Intensity and Turbulence Convection Velocity Measurements for a Convergent Nozzle in a Free Jet Wind Tunnel*; NASA Contractor Report 2949, Contract NAS3-17866; NASA: Washington, DC, USA, 1978.

© 2017 by the authors. Licensee MDPI, Basel, Switzerland. This article is an open access article distributed under the terms and conditions of the Creative Commons Attribution (CC BY) license (http://creativecommons.org/licenses/by/4.0/).

applied sciences

MDPI

Article

Simulation of Tail Boom Vibrations Using Main Rotor-Fuselage Computational Fluid Dynamics (CFD)

Andrey Batrakov [1], Alexander Kusyumov [1,*], Sergey Kusyumov [1], Sergey Mikhailov [1] and George N. Barakos [2]

1 Aerohydrodynamics Department, Kazan National Research Technical University n.a. A.N. Tupolev (KNRTU-KAI), 10 Karl Marx St., Kazan 420111, Russia; batrakov_a.c@mail.ru (A.B.); kusok88@yandex.ru (S.K.); sergey.mikhaylov@kai.ru (S.M.)

2 School of Engineering, University of Glasgow, Glasgow G12 8QQ, UK; george.barakos@glasgow.ac.uk

* Correspondence: postbox7@mail.ru; Tel.: +7-917-264-8584

Received: 25 July 2017; Accepted: 1 September 2017; Published: 7 September 2017

Abstract: In this work, fully-resolved rotor-fuselage interactional aerodynamics is used as the forcing term in a model based on the Euler-Bernoulli equation, aiming to simulate helicopter tail-boom vibration. The model is based on linear beam analysis and captures the effect of the blade-passing as well as the effect of the changing force direction on the boom. The Computational Fluid Dynamics (CFD) results were obtained using a well-validated helicopter simulation tool. Results for the tail-boom vibration are not validated due to lack of experimental data, but were obtained using an established analytical approach and serve to demonstrate the strong effect of aerodynamics on tail-boom aeroelastic behavior.

Keywords: Computational Fluid Dynamics (CFD); helicopter main rotor; tail-boom vibration; Euler-Bernoulli equation

1. Introduction

Interactional effects between the main rotor and the fuselage are commonplace in rotorcraft. In particular, at low advance ratios of the main rotor, its wake can interact with the main fuselage and the empennage. The higher the helicopter weight, the stronger the rotor wake, and therefore interactional aerodynamics can be significant. Leishman [1,2] carried out experiments on the topic and there are several numerical studies too. Nevertheless, amongst the current research reports there is little information on the effect of these aerodynamic interactions on the vibration and deformation of the tail-boom structure. This interaction is expected to be significant if a long tail-boom is used or if the employed structure is light with substantial weight added at the end of the boom due to the presence of the tail rotor, its rotor-head, the intermediate gear-box, fin and horizontal stabilizer. One can expect that the tail boom loads are not high, but estimating the vibration level is important for design and fatigue analysis of the boom. Therefore, this work aims to investigate the effect of unsteady aerodynamics on the vibration and deformation of a helicopter tail-boom. In contrast to earlier works [3] where the rotor aerodynamics is represented by blade-element methods, Computational Fluid Dynamics (CFD) is used here to compute the unsteady flow, and then the aerodynamic loads are used in conjunction with a simple structural model. The use of CFD allows for the details of the unsteady flow spectrum to be captured and preserved in the flow around the tail-boom.

In terms of modeling the structure of the tail-boom, it was decided here to keep the model simple, based on the Euler-Bernoulli equation for coupling with CFD via embedding the model in the framework of a CFD solver. The Euler-Bernoulli equation for various end conditions allows for analytical or approximate solution of the natural and forced vibrations of uniform and non-uniform beams [4–9].

Detailed empennage models with structural optimization were considered in [10], where complete Finite Element Method (FEM) models are demonstrated for computing the natural mode shapes and frequencies of the structure. In [11], maintaining the non-dimensional amplitude of the forces against several lifting conditions were analyzed with the FEM approach. Nevertheless, these works did not proceed to compute the effects of wake aerodynamics on the tail boom.

To determine the transverse vibrations of an Euler-Bernoulli uniform beam in the present study, an approximate analytical approach is used, based on the approach of [12]. According to [12], a solution of the Euler-Bernoulli can be presented as a series of spatial and time coordinates. The forced vibrations are computed using an approximation of the tail boom load obtained from CFD calculation of rotor-fuselage configuration.

The selected test case is motivated by the Ansat light helicopter (Kazan Helicopters Public Stock Company, Kazan, Russia), although the results obtained serve to demonstrate the employed method and correspond to a generic interaction case. The aerodynamic predictions of the helicopter fuselage aerodynamics are validated against wind tunnel tests.

2. Fuselage Aerodynamics

The first step of this work is the simulation of the flow around the isolated "clean" fuselage. The Helicopter Multi-Block (HMB) solver is established in the field of rotorcraft aerodynamics and it is based on the discretization of the Unsteady Reynolds Averaged Navier-Stokes (URANS) equation on multi-block structured grids. The solver allows for sliding and overset grids and has a variety of solution methods for flows at low or very high Mach number. A mesh deformation method based on a combination of the Trans-Finite Interpolation and the Spring-Analogy method allows for aero-elastic cases to be studied. Also, a sliding-mesh method is implemented so that test cases with relative motions of different parts of the geometry can be modeled. The HMB method has been validated for a range of rotorcraft applications [13–18] and has demonstrated good accuracy and efficiency for very demanding flows. The parallel implementation makes use of the Message Passing Interface library for inter-processor communication and of parallel I/O for saving and reading data from out-of-core storage. The HMB method has so far been used for the analysis of rotors, wind turbines, propellers and cavities and has demonstrated good scalability for up to 10 k cores. This was of course achieved using a fine mesh. A summary of the method in HMB is presented in reference [13]. The code can also use actuator disks or virtual blade models to simulate the effect of the main rotor on the fuselage.

Although the solver is able to use Detached-Eddy Simulation (DES) and Large-Eddy Simulation (LES) models, the URANS equations were used in this work. This was justified by the rather limited regions of flow separation encountered during computations. In general, different models of turbulence, including the Spalart-Allmaras one-equation model, the k-ω (SST) 2-equation model and transition models of turbulence are available in HMB solver.

The geometry of the isolated fuselage corresponds to an early Ansat-P model of the Ansat helicopter produced by the Kazan Helicopters. The wind tunnel model (Figure 1a) was manufactured based on the computer model, used for CFD modeling. The aerodynamic analysis of this model was considered in references [19,20], and were studied using the open test section (2.25 m diameter) closed circuit, low speed, wind tunnel T-1K of the KNRTU-KAI Aerohydrodynamics department.

For the grid around the Ansat-P fuselage, the (ICEM)-hexa software of ANSYS^{TM} mesh generation tool has been used. The length of the wind tunnel model (Figure 1a) was L_F = 1.8 m. The computational grid for this model contained 964 blocks and 13.5×10^6 cells. The surface grid and grid details are shown in Figure 1b–d.

The topology and surface grid near the area of the engine exhausts is also presented on the same figure. Care has been taken to represent the geometry of the wind tunnel models as accurately as possible, regardless of the minor edits that the laser-scans of the models needed in order to be converted to air-tight surfaces suitable for CFD.

In terms of turbulence modeling, the k-ω model [21] was employed. Experimental and CFD analyses were conducted at Reynolds number of 4.4×10^6 and Mach number of 0.1.

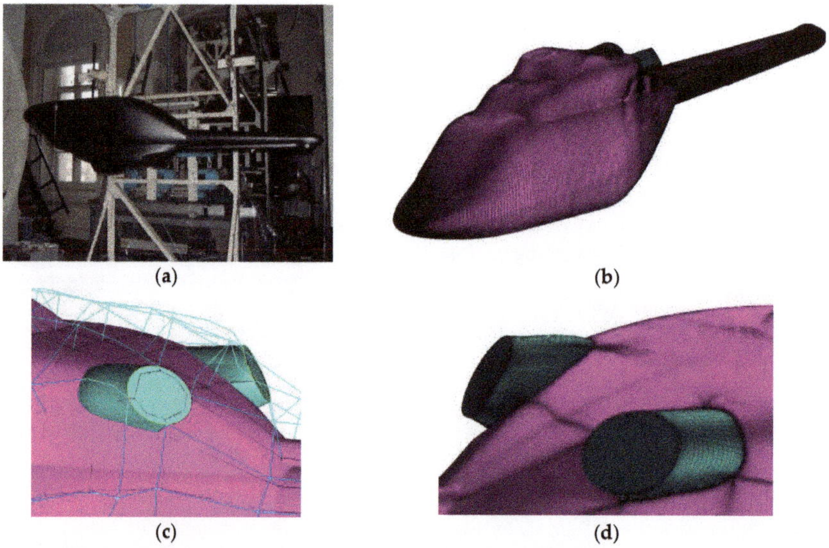

Figure 1. (**a**) Ansat-P fuselage model in the T-1K wind tunnel of KNRTU-KAI, (**b**) surface grid for fuselage, (**c**) multi-block topology, and (**d**) surface mesh near exhausts.

Figure 2 suggests a good agreement between CFD and experimental results for the lift (C_L) and drag (C_D) coefficients in the considered range of pitch angles.

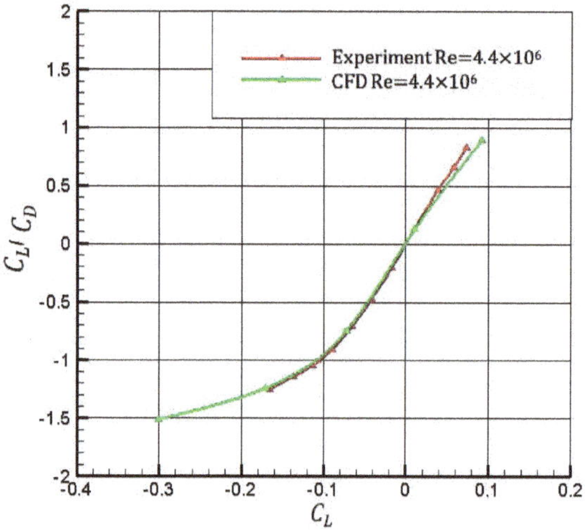

Figure 2. Computational Fluid Dynamics (CFD) and experimental drag coefficients vs. lift coefficient for Ansat-P model.

More detailed information about grid sensitivity studies and the HMB code validation vs. experimental data is presented in [14,19,20].

The next step was the simulation of the flow around the rotor model in hover [22]. The CFD code validation was performed against the experimental data of Caradonna and Tung [23] that are extensively used within the helicopter community. Figure 3 shows the satisfactory agreement of CFD results for the pressure coefficient (Cp) distribution with experimental data ($M_{tip} = 0.612$, collective pitch of 8 degrees) at two values of the rotor dimensionless radius $\bar{r}$.

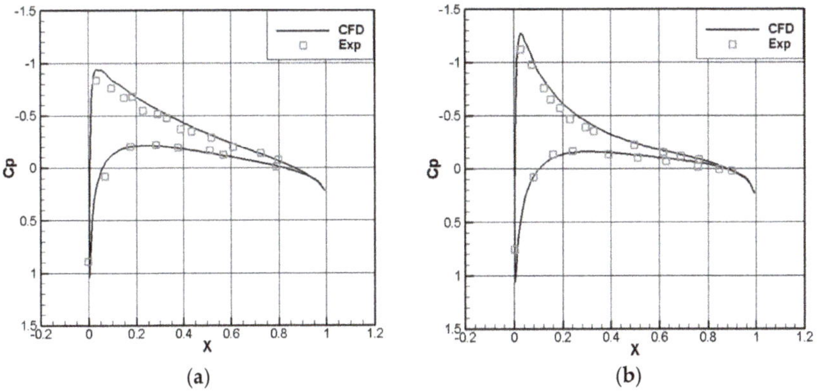

Figure 3. Pressure coefficients along the rotor section: (**a**) $\bar{r}$ = 0.68, (**b**) $\bar{r}$ = 0.96.

3. Rotor-Fuselage Computations

Rotor-fuselage flow simulations are also considered in this paper. For the simulation of the rotor over the fuselage, a sliding surface is constructed that divides the computational domain into two parts (Figure 4). The movable upper part corresponds to the rotor, that includes four-blades made of NACA 23012 airfoils with the root cut-off $r = 0.2R$, where R is the radius of the rotor. A simplified elliptical hub is used. The geometry of the upper part is inclined to allow forward tilting of the rotor.

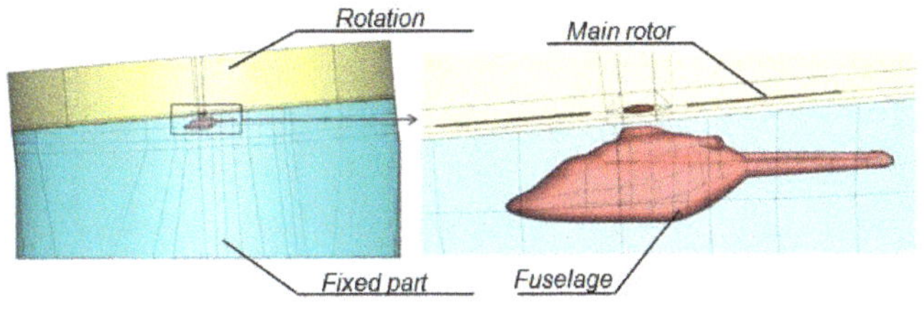

Figure 4. Far and near view of the computation domain.

The CFD grids are constructed using the ICEM-hexa software of ANSYS^{TM}. The topology of the blocks and the parameters of the computational grids correspond to what was used for the isolated fuselage of the helicopter. The fixed part of the mesh contains 688 blocks and 9×10^6 cells.

The computational grid of the rotor was assembled in several stages. At the first stage, the computational grid for a quarter of the computational domain (for one blade) was generated, as shown in Figure 5. The computational grid for each blade comprises 144 blocks and 6×10^6 cells.

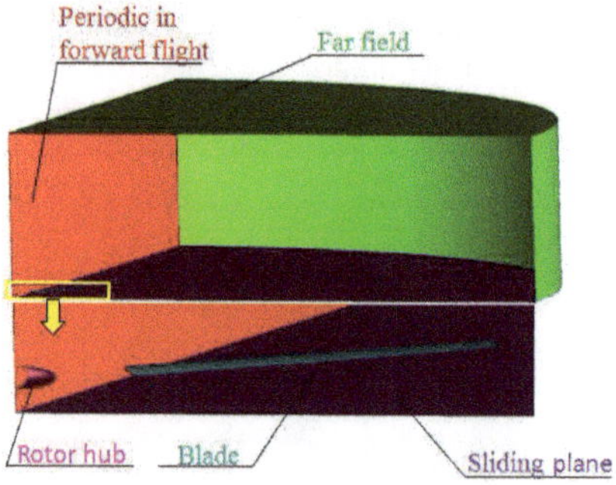

Figure 5. Sliding plane arrangement near the main rotor hub.

The boundary conditions are shown in Figure 5. Only a single blade was meshed, but using periodic conditions between blades allows for the full rotor to be considered by copying and rotating the mesh of a single blade around the azimuth (Figure 6). This method simplifies the process of constructing the computational grid for the main rotor and keeps constant the characteristics of the computational grid for each blade.

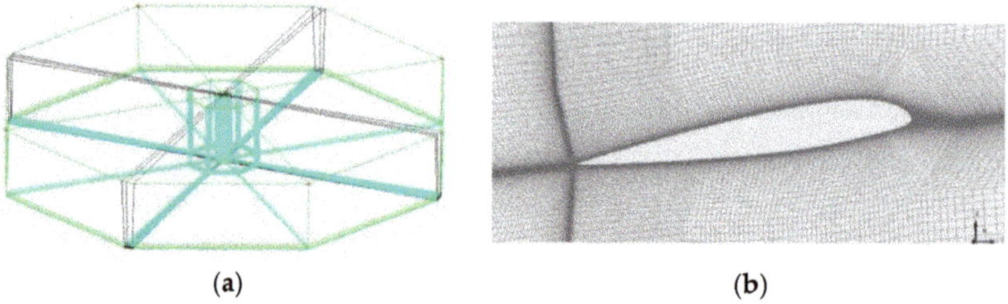

Figure 6. Multi-block topology (**a**) and mesh section (**b**).

After assembling of all elements, the computational grid had 1144 blocks and 33×10^6 cells. The simulation concerns forward flight for a 1:6 scaled helicopter model, and all geometric and flight parameters are presented in Table 1.

Table 1. Main rotor parameters.

Geometry Parameters	
Number of blades, N	4
Rotor diameter, $2R$ (m)	1.92
Root cut-out, (m)	0.19
Blade twist, ϕ (deg)	−5.3
Blade chord, c (mm)	52
Blade thickness, f (%c)	12
Operation Parameters	
Collective pitch angle, θ_0 (deg)	8
Cyclic pitch angle, θ_{1s} (deg)	−2
Cyclic pitch angle, θ_{1c} (deg)	2
Coning angle, β (deg)	0
Angle of attack, α (deg)	−4
Tip Mach number	$M_{tip} = 0.64$
Advance ratio, μ	0.15
Direction of rotation	Counter clockwise

The following assumptions were made:

- Rigid blades;
- No flapping motion of the blades, only pitch input is considered;
- No lead-lag.

Changing the cyclic pitch of the rotor is achieved by deforming the computational grid. The employed method is described in [13].

The simulation was unsteady, and with a time step corresponding to 1 degree in rotor azimuth. An example of the surface pressure distribution at the azimuth of $\psi = 80°$ is shown in Figure 7.

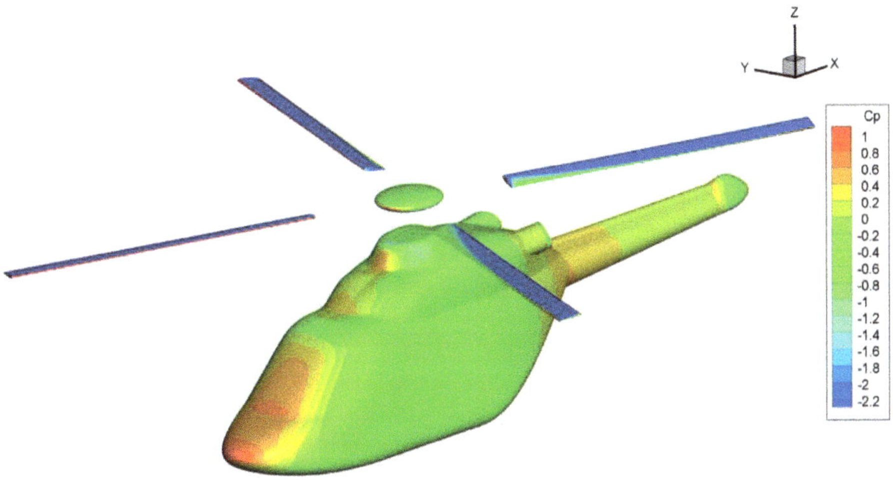

Figure 7. Surface pressure coefficient on the fuselage and blades at the conditions of Table 1.

Analysis of the results was carried out using Tecplot 360™, and normal vectors to the surface of the body oriented towards the outside were computed.

Variation of the pressure distribution over the surface of the fuselage and the rotor blades leads to variable fuselage drag (C_D) and rotor thrust (C_T) coefficients, which are computed by:

$$C_D = \frac{D}{q_\infty S_F}, C_T = \frac{T}{q_{tip} \pi R^2},$$

where q_∞ is the free stream dynamic pressure, q_{tip} is the dynamic pressure at the blade tip, S_F is the reference fuselage area, D is the drag force, and T the rotor thrust. The oscillation amplitude values of the coefficient C_D of drag of the fuselage are about 13% around the mean (Figure 8).

The average value of drag in the presence of a fuselage of the rotor is higher compared with the simulation results of an isolated flow fuselage. Increased drag of 46.6% is seen, which correlates with the results obtained by using a simplified actuator-disk model [24] (ΔC_D = 45.6% at the value of thrust coefficient C_T = 0.0128).

The value of the thrust coefficient also changes during the rotation of the rotor (Figure 9). The oscillation amplitude reaches 2.25% of the mean.

The rotation of the rotor has a strong effect on the fuselage. Figure 10 presents diagrams of loading of the tail boom at different azimuthal positions of the rotor.

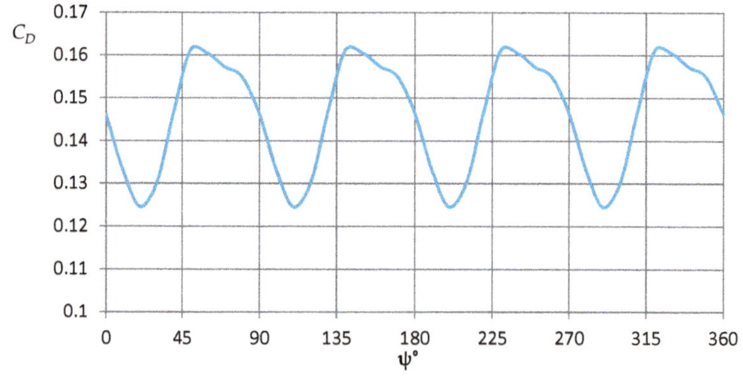

Figure 8. Fuselage drag as function of the blade azimuth.

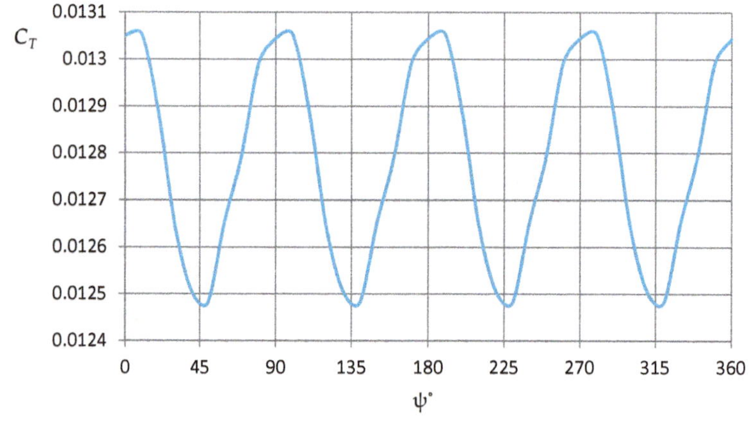

Figure 9. Main rotor thrust as function of the blade azimuth.

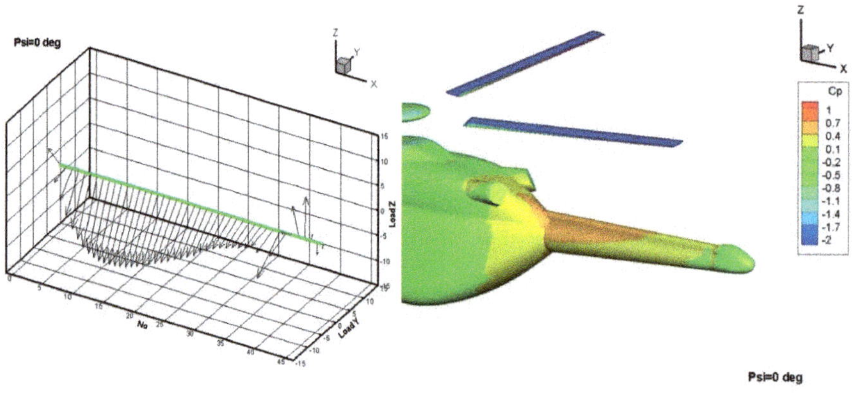

(a)

Figure 10. *Cont.*

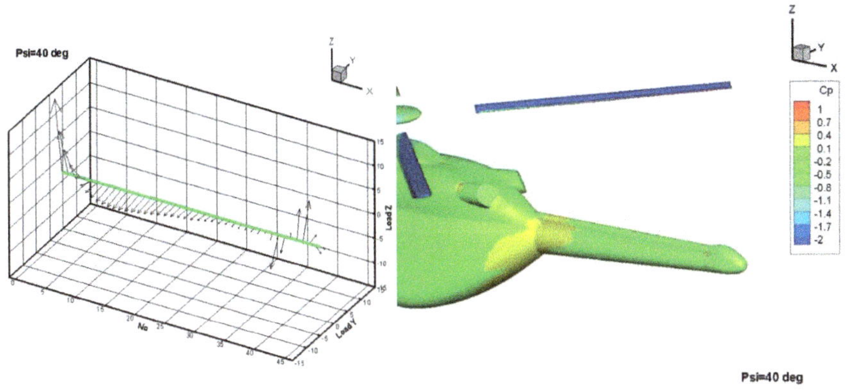

(**b**)

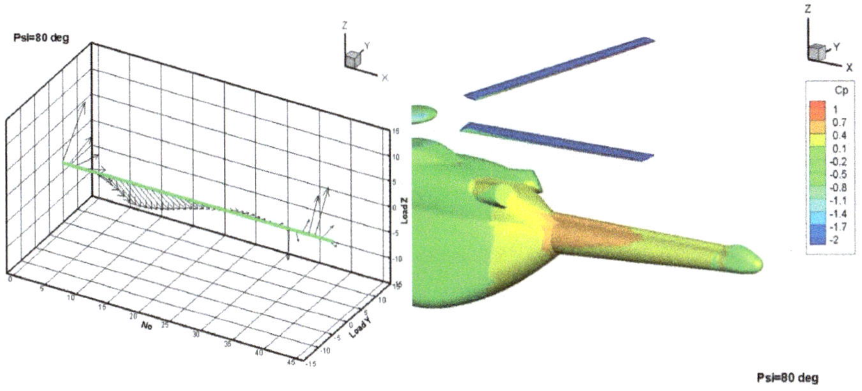

(**c**)

Figure 10. Load Z and Load Y present sectional forces along the tail boom projected in the vertical (Z) and lateral directions for different azimuthal positions: (**a**) $\psi = 0°$, (**b**) $\psi = 40°$, (**c**) $\psi = 80°$.

The aerodynamic load acting on the tail boom is shown in Figure 11 in terms of vertical $F_{Bz}(\psi)$ force coefficient that was computed according to the expression:

$$C_{Bz}(\psi) = \frac{F_{Bz}(\psi)}{q_\infty S_F} \quad (1)$$

Figure 11 shows that the effect of the rotor on the tail boom is characterized by rapid loading changes. The main oscillation frequency corresponds to the blade passing frequency.

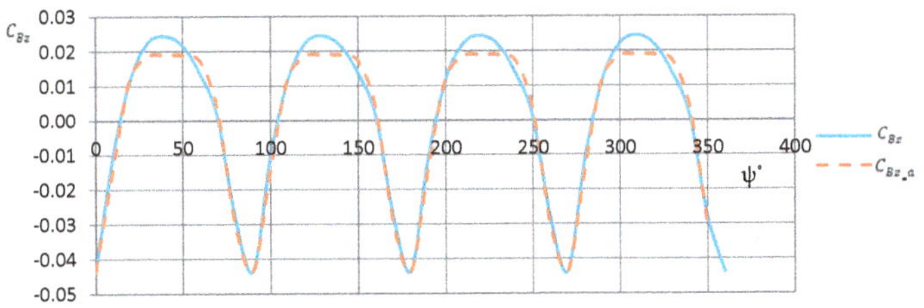

Figure 11. Vertical aerodynamic load on the tail boom, as a function of the azimuth of the main rotor blades.

The variation of $C_{Bz}(\psi)$ can be approximated by a function

$$C_{Bz\_a}(\psi) = \alpha_1(1 + \alpha_2[1 + \alpha_3 \cos(\gamma + N\psi)]^N) \tag{2}$$

where $\alpha_1 = 0.0189$, $\alpha_2 = -0.329$, $\alpha_3 = 0.7891$, $\gamma = 0.15$ rad ($N = 4$ is the number of blades). Using a harmonic function (cosine) in expression (2) captures the periodic character of the tail-boom load. The parameter α_1 allows for correction of the peak to peak amplitude and the parameters α_2, α_3 determine the shift of the $C_{Bz\_a}(\psi)$ function with respect to the azimuth axis. The parameters α_i were fitted to ensure $\int_0^{2\pi} C_{Bz\_a}(\psi)d\psi \approx \int_0^{2\pi} C_{Bz}(\psi)d\psi$. The function $C_{Bz\_a}(\psi)$ is shown in Figure 11 in comparison to the function $C_{Bz}(\psi)$.

4. Mathematical Model of Tail Boom Vibrations

The tail boom of the helicopter is susceptible to vibration. Several factors contribute to this; in particular, the impact of the main and tail rotor loads. An approximate analytical method for simulation of the tail boom vibrations is considered here.

The mathematical formulation of the problem is bound by the following limitations and assumptions: (1) The tail boom is considered to be of a constant diameter thin-walled cylindrical structure with continuously distributed mass (no concentrated mass points), so that the total mass of the beam is equal to the one of the light helicopter boom (but also adding the horizontal tail, transmission and other design elements); (2) The tail boom is rigidly fixed at the fuselage end and the other end is free; (3) The mathematical model of vibrations (vertically directed) without viscous damping is determined by the Euler-Bernoulli equation with one spatial coordinate and variable tail boom geometry along the tail boom span.

Under these assumptions, the equation of the tail boom deformation is described by equation [4,7,12]:

$$\frac{\partial^2}{\partial x^2}\left\{EI\frac{\partial^2 v}{\partial x^2}\right\} + m_L\frac{\partial^2 v}{\partial t^2} = F_L(t,x). \tag{3}$$

Here x is the longitudinal coordinate; t is the time coordinate; EI is the flexural rigidity (E is Young's modulus, I is a moment of inertia); $v(t,x)$ is the transverse vertical deformation; m_L is the mass per unit length. The normal (vertical) component $F_L(t,x)$ of the force acting on the surface of the tail boom per unit length is given by

$$F_L(t,x) = \frac{F_{Bz}(t,x)}{L} = q_\infty\frac{S_F}{L}c_F(t,x). \tag{4}$$

Here $c_F(t,x)$ is the normal force coefficient; L is length of the beam. The boundary conditions for the Equation (3) can be written as

$$\begin{gathered} v(0,x) = \varphi(x);\ \frac{\partial v}{\partial t}(0,x) = \xi(x); \\ v(t,0) = 0;\ \frac{\partial v}{\partial x}(t,0) = 0; \\ \frac{\partial^2 v}{\partial x^2}(t,L) = 0;\ \frac{\partial^3 v}{\partial x^3}(t,L) = 0. \end{gathered} \tag{5}$$

According to [12], the solution of Equation (3) with boundary conditions (5) can be written in the form

$$v(t,x) = v_0(t,x) + v_1(t,x). \tag{6}$$

Here the functions $v_0(t,x)$ and $v_1(t,x)$ can be determined by using constraints of the physical and mathematical formulation of the vibration task: the function $v_0(t,x)$ determines the natural and the function $v_1(t,x)$ determines forced tail boom vibrations. Substitution of (6) in Equation (3) leads to the expression:

$$b^2\frac{\partial^2}{\partial x^2}\left\{\left(\frac{\partial^2 v_0}{\partial x^2}+\frac{\partial^2 v_1}{\partial x^2}\right)\right\}+\left(\frac{\partial^2 v_0}{\partial t^2}+\frac{\partial^2 v_1}{\partial t^2}\right)=F_{Lm}(t,x). \tag{7}$$

Here $F_{Lm}(t,x)=F_L(t,x)/m_L$, $b^2=EI/m_L$.

From the last expression, a system of two equations can be obtained:

$$b^2\frac{\partial^4 v_0}{\partial x^4}+\frac{\partial^2 v_0}{\partial t^2}=0 \tag{8}$$

$$b^2\frac{\partial^4 v_1}{\partial x^4}+\frac{\partial^2 v_1}{\partial t^2}=F_{Lm}(t,x). \tag{9}$$

Boundary conditions for the Equation (8) can be written as

$$\begin{gathered}v_0(t,0)=0,\ \frac{\partial v_0}{\partial x}(t,0)=0,\\ \frac{\partial^2 v_0}{\partial x^2}(t,L)=0,\ \frac{\partial^3 v_0}{\partial x^3}(t,L)=0,\\ v_0(0,x)=\varphi(x),\ \frac{\partial v_0}{\partial t}(0,x)=\xi(x).\end{gathered} \tag{10}$$

For Equation (9), the boundary conditions are taken in the form

$$v_1(0,x)=0,\ \frac{\partial v_1}{\partial t}(0,x)=0.$$

The Equation (8) can be rewritten in a dimensionless form

$$\frac{\partial^4 \overline{v}_0}{\partial \overline{x}^4}+\frac{\partial^2 \overline{v}_0}{\partial \overline{t}^2}=0, \tag{11}$$

with boundary conditions

$$\begin{gathered}\overline{v}_0(\overline{t},0)=0,\ \frac{\partial \overline{v}_0}{\partial \overline{x}}(\overline{t},0)=0,\\ \frac{\partial^2 \overline{v}_0}{\partial \overline{x}^2}(\overline{t},1)=0,\ \frac{\partial^3 \overline{v}_0}{\partial \overline{x}^3}(\overline{t},1)=0,\\ \overline{v}_0(0,\overline{x})=\overline{\varphi}(\overline{x}),\ \frac{\partial \overline{v}_0}{\partial \overline{t}}(0,\overline{x})=\overline{\xi}(\overline{x}).\end{gathered} \tag{12}$$

Similarly, Equation (9) can be written as:

$$\frac{\partial^4 \overline{v}_1}{\partial \overline{x}^4}+\frac{\partial^2 \overline{v}_1}{\partial \overline{t}^2}=\overline{F}_{Lm}(\overline{t},\overline{x}) \tag{13}$$

with boundary conditions

$$\overline{v}_1(0,\overline{x})=0,\ \frac{\partial \overline{v}_1}{\partial \overline{t}}(0,\overline{x})=0. \tag{14}$$

The dimensionless variables in (13) and (14) are determined by the expressions

$$\overline{v}_1=\frac{v_1}{L},\overline{v}_0=\frac{v_0}{L},\overline{t}=\frac{t}{t_0},\ \overline{x}=\frac{x}{L},$$

where $t_0 = L^2/b$ is a reference time. The right part of (13) according to (4) can be written as

$$\overline{F}_{Lm}(\overline{t},\overline{x}) = F_{Lm}(t_0\overline{t}, L\overline{x})\frac{t_0^2}{L} = \overline{F}_0 c_F(\overline{t},\overline{x}). \quad (15)$$

Here $c_F(\overline{t},\overline{x})$ is a normalized transversal load coefficient, and $\overline{F}_0 = q_\infty S_F t_0^2/(m_L L^2)$ is a constant. According to [4,7,12], the solution of Equation (11) can be presented in the form

$$\overline{v}_0(\overline{t},\overline{x}) = \sum_{n=1}^{\infty} T_n(\overline{t})X_n(\overline{x}), \quad (16)$$

where $T_n(\overline{t}),\ X_n(\overline{x})$ are dimensionless functions. Substitution of (16) in (11) yields the system of equations

$$T_n''(\overline{a}_n)^4 T_n = 0,\ X_n - (\overline{a}_n)^{-4} X_n^{IV} = 0. \quad (17)$$

The coefficients $\overline{a}_n = a_n L$ are determined by the relations [4,7,12]:

$$\overline{a}_1 = 1.875, \overline{a}_2 = 4.694,\ \overline{a}_3 = 7.855,\ \ \overline{a}_n = \pi(2n-1)/2$$

where $n = 4, 5, 6, 7, \ldots$

The solution of the system (17) with the boundary conditions of (12) can be written as

$$X_n(\overline{x}) = \left[\cos(\overline{a}_n\overline{x}) - \cosh(\overline{a}_n\overline{x}) - \frac{\cos(\overline{a}_n) + \cosh(\overline{a}_n)}{\sin(\overline{a}_n) + \sinh(\overline{a}_n)}(\sin(\overline{a}_n\overline{x}) - \sinh(\overline{a}_n\overline{x}))\right]$$

$$T_n = A_n\cos(\overline{\omega}_n\overline{t}) + B_n sin(\overline{\omega}_n\overline{t})$$

Here,

$$A_n = \int_0^1 \overline{\varphi}(\overline{x})X_n(\overline{x})d\overline{x},\ B_n = 1/(\overline{\omega}_n)\int_0^1 \overline{\xi}(\overline{x})X_n(\overline{x})d\overline{x},$$

and

$$\overline{\omega}_n = \overline{a}_n^2 = \omega_n t_0 \quad (18)$$

So, the solution of Equation (11) has the form [4,7,12]

$$\overline{v}_0(\overline{t},\overline{x}) = \sum_{i=1}^{\infty} A_n X_n(\overline{x})\cos(\overline{\omega}_n\overline{t}) + \sum_{i=1}^{\infty} B_n X_n(\overline{x}) sin(\overline{\omega}_n\overline{t}) \quad (19)$$

Figure 12 shows the basic functions $X_n(\overline{x})$ $(n = 1,\ldots,4)$ for the simulated conditions. One can note that the shape of calculated $X_n(\overline{x})$ functions corresponds to the reference data (see, for example [4]).

The solution of Equation (13) can be written as [12]

$$\overline{v}_1(\overline{t},\overline{x}) = \sum_{n=1}^{\infty} S_n(\overline{t})X_n(\overline{x}) \quad (20)$$

where the function $\overline{R}(\overline{t},\overline{x})$ is approximated by the series

$$\overline{F}_{Lm}(\overline{t},\overline{x}) = \sum_{n=1}^{\infty} H_n(\overline{t})X_n(\overline{x})$$

Substituting (20) in (13) and after some transformations result in

$$\sum_{n=1}^{\infty}\left(S_n'' + \overline{\omega}_n^2 S_n\right)X_n = H_n X_n$$

that is a system of equations

$$S_n'' + \overline{\omega}_n^2 S_n = H_n$$

with boundary conditions

$$S_n(0) = 0,\ S_n'(0) = 0.$$

The functions $S_n(\overline{t})$ and $H_n(\overline{t})$ are determined by the expressions:

$$H_n(\overline{t}) = \int_0^1 \overline{F}_{Lm}(\overline{t},\overline{x})X_n(\overline{x})d\overline{x}, \tag{21}$$

$$S_n(\overline{t}) = \frac{1}{\overline{\omega}_n}\int_0^{\overline{t}} H_n(\sigma)sin\left[\overline{\omega}_n(\overline{t}-\sigma)\right]d\sigma.$$

Thus, based on the functions $\overline{v}_0(\overline{t},\overline{x})$ and $\overline{v}_1(\overline{t},\overline{x})$, the general solution of the Equation (3) can be presented in the form of (6), (16) and (20) with a finite number of terms of the series expansion.

From the solution above, it follows that, in this study, the normalized functions $\overline{v}_0(\overline{t},\overline{x})$ and $\overline{v}_1(\overline{t},\overline{x})$ determine the natural and forced vibrations, respectively, for the equivalent uniform beam.

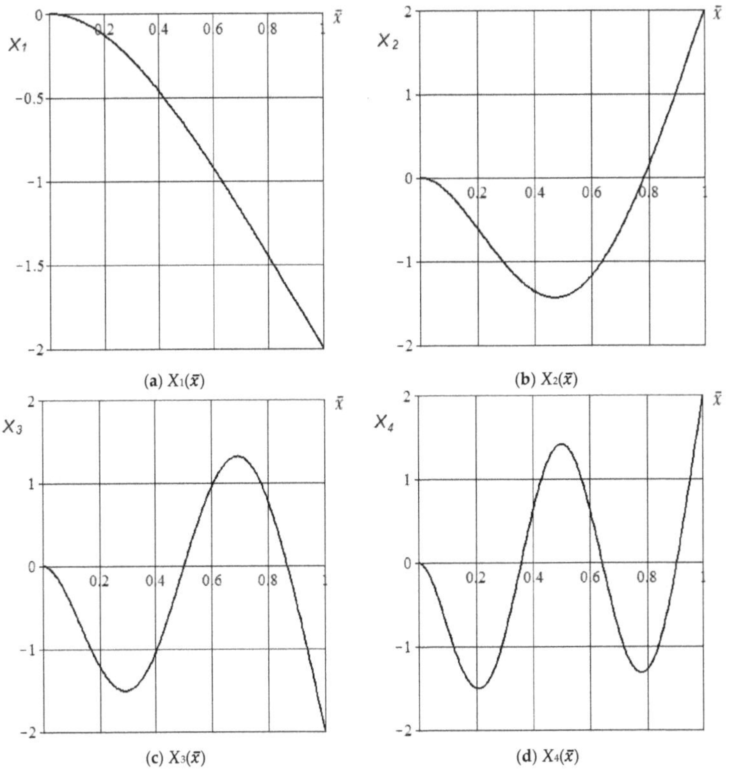

Figure 12. The basic functions $X_n(\overline{x})$.

5. Calculation of the Tail Boom Vibrations

The geometric parameters and physical tail boom material properties are presented in Table 2 and approximately correspond to parameters of a full scale light helicopter.

Table 2. Tail boom parameters.

Parameters	
Diameter of the fixed beam end, D_1 (m)	0.546
Diameter of the free beam end, D_2 (m)	0.346
Beam length, L (m)	4
Wall thickness of the beam, δ (m)	0.001
Thickness of the stringer, δ_S (m)	0.003
Length of the stringer, L_S (m)	0.015
Number of stringers, N_S	10
Beam material density, ρ_b (kg/m^3)	2.7×10^3
Young's modulus, E (MPa)	72×10^3

The formulation adopted in this paper does not account for a non-uniform tail boom surface. For this reason, computations of forced vibrations were conducted for several equivalent uniform tail booms with diameter and mass per unit length, obtained from:

$$D = \frac{D_1 + D_2}{2}(1+\gamma), m_L = \rho_b(N_s L_s \delta_s + 2\pi\delta D) + m_c,$$

where γ is a small parameter, $m_c = 30$ kg/m is a mass of internal tail boom construction elements per unit length. Table 3 presents the values of the γ parameters, mass and geometry used for forced vibration simulations.

The beam moment of inertia is determined, taking into account properties of the stringers used typically for such beams. The moment of inertia of the stringers is determined by the expression:

$$I_s = \frac{N_s}{8} D^2 \delta_s L_s,$$

so that the total moment of inertia of the tail boom is

$$I = \frac{\pi}{64}\left(D^4 - (D-\delta)^4\right) + I_s.$$

Based on the results of the CFD simulation of the rotor-fuselage interaction, the normal force coefficient was determined in the form $c_F(\bar{t},\bar{x}) = c_{Ft}(\bar{t})c_{Fx}(\bar{x})$, where $c_{Ft}(\bar{t}),\ c_{Fx}(\bar{x})$ are trigonometric functions. From Figure 10, it follows that $c_{Fx}(0) \approx 0$, $c_{Fx}(1) \approx 0$, $|c_{Fx}(0.5)| \approx max$. The results of Figure 11 and the function $c_{Bz\_a}(\psi)$ (expression (2)) were used to determine the function $c_{Ft}(\bar{t})$. In this case, the function $c_{Fx}(\bar{x})$ has to satisfy to the condition $\int_0^1 c_{Fx}(\bar{x})d\bar{x} = 1$. So, an approximation of the function $c_F(\bar{t},\bar{x})$ in (15) was taken here as

$$c_F(\bar{t},\bar{x}) = 0.5\pi\alpha_1 \sin(\pi\bar{x})\left(1 + \alpha_2\left[1 + \alpha_3\cos(\gamma + N\bar{\omega}\bar{t})\right]^N\right)$$

where $f = 6.023$ Hz is the main rotor frequency, $\omega = 2\pi f$ is the angular rotor frequency, and $\bar{\omega} = \omega t_0$.

Table 3. Equivalent mass and diameter of tail boom.

Variant, N	Parameter γ	Diameter D, m	Mass Per Unit Length m_L, (kg/m)	Dimensionless Rotor Frequency $N\bar{\omega}$
1	0.224	0.546	40.48	9.07
2	0	0.446	38.78	10.52
3	−0.02914	0.433	38.56	10.9
4	−0.03812	0.4293	38.498	11.0168
5	−0.04484	0.426	38.44	11.12
6	−0.06502	0.417	38.29	11.41
7	−0.225	0.346	37.08	14.27

One should note here that, for the employed main rotor frequency of 6.023 Hz, the time step Δt corresponding to 1 degree in rotor azimuth is 612×10^{-4} s, that satisfies the condition $(\Delta t)^{-1} \gg f_1$. The peak value of the normal force coefficient $\left|c_{Ft}(\bar{t})\right| \approx 0.043$ for the considered simulation parameters corresponds to a peak vertical tail boom load of about 30 N per meter of tail boom length.

Using the basic functions $X_n(\bar{x})$, one can determine the function $\bar{v}_1(\bar{t},\bar{x})$ for the forced tail boom vibrations (analytical transformations and calculations were performed using Maple 17™ software).

One can note here that the behavior of the $\bar{v}_1(\bar{t},1)$ function is determined primarily by the first two components $S_1(\bar{t})X_1(\bar{x})$ and $S_2(\bar{t})X_2(\bar{x})$. The eigenfrequencies f_1 and f_2 depend on the geometry parameter γ. Nevertheless, in this work, all components $S_n(\bar{t})X_n(\bar{x})$, $n = 1,\ldots,4$ were used to determine $\bar{v}_1(\bar{t},\bar{x})$. Thus, a general behavior of the $\bar{v}_1(\bar{t},1)$ function depends on the interaction of the eigen and forced vibrations. Figure 13 shows the functions $S_1(\bar{t})$ and $S_2(\bar{t})$ for the different values of parameter γ. The time behavior of the functions $S_1(\bar{t})$ and $S_2(\bar{t})$ is predominantly determined by the frequencies $\bar{\omega}_1$ and $\bar{\omega}_2$, respectively, and the amplitude of the total function $\bar{v}_1(\bar{t},\bar{x})$ depends on the mass and geometry data. For variant 4, the function $S_2(\bar{t})$ demonstrates resonance due to the second blade passing harmonic $2Nf$ being equal to the second eigenfrequency f_2 and $2N\bar{\omega} = \bar{\omega}_2 = 22.0336$.

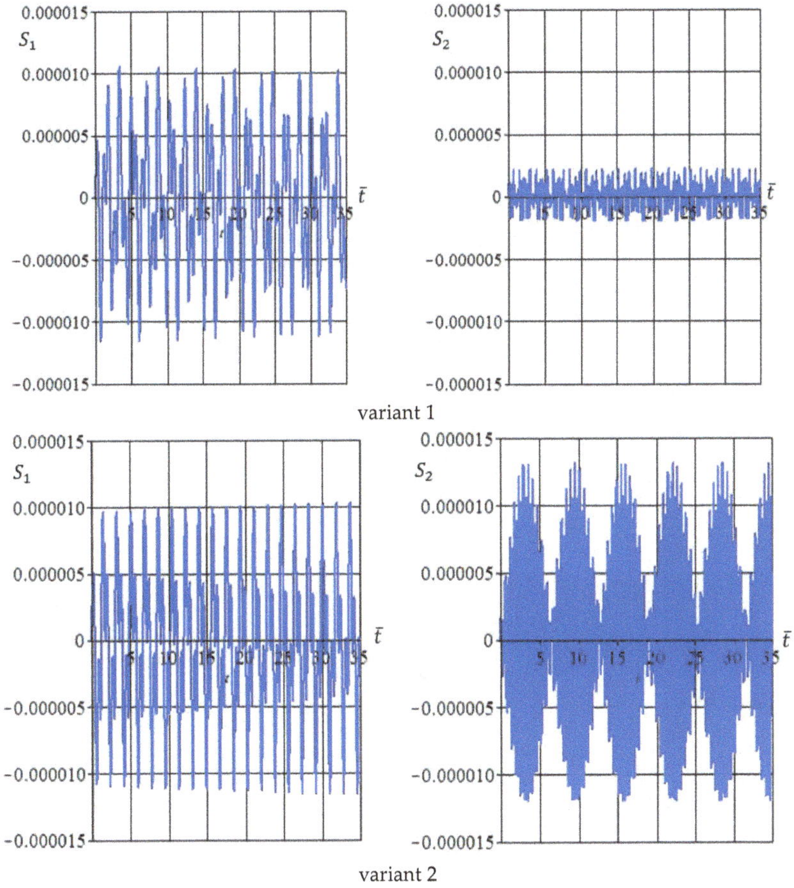

Figure 13. *Cont.*

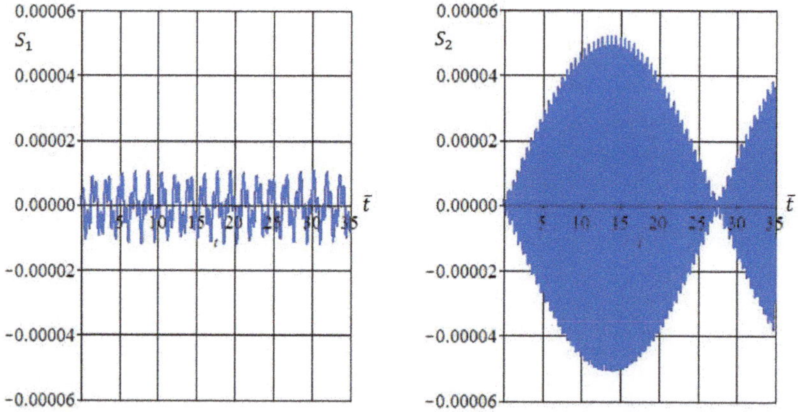

variant 3

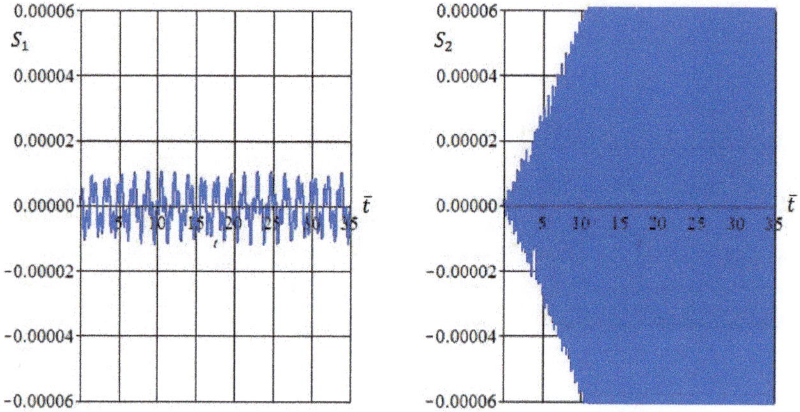

variant 4

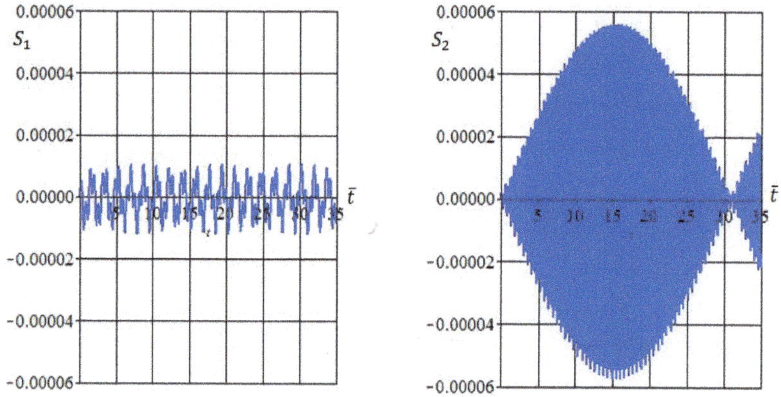

variant 5

Figure 13. *Cont.*

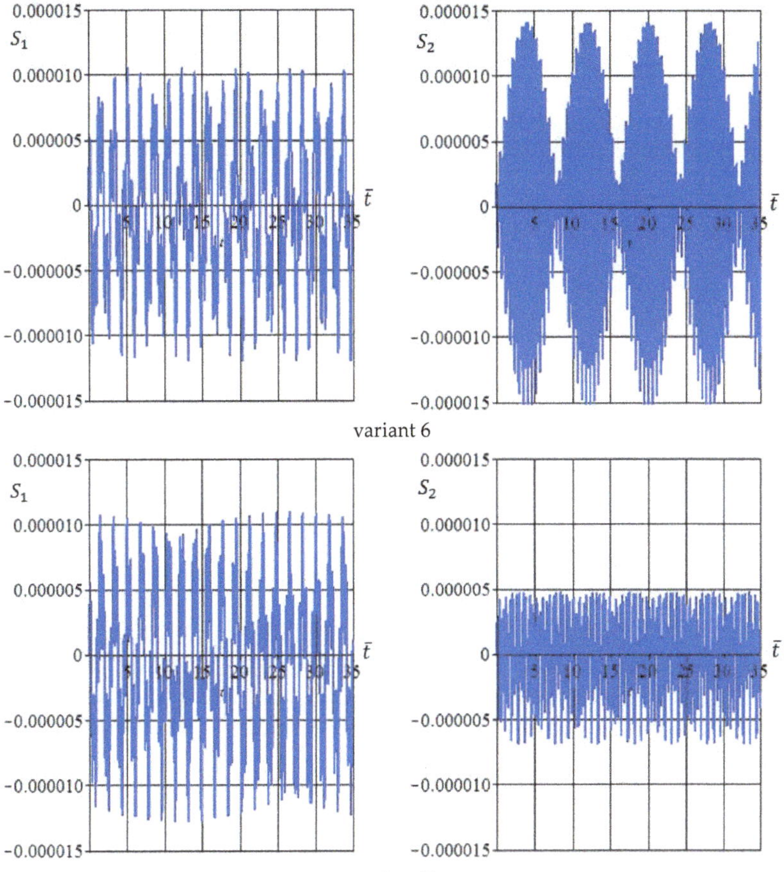

Figure 13. Functions $S_1(\bar{t})$ (**left column**) and $S_2(\bar{t})$ (**right column**) for different γ values (variant numbers).

From Figure 13, it follows that, for variants 1 and 7, the amplitude of the forced tail boom oscillations is determined by the $S_1(\bar{t})X_1(\bar{x})$ term and one can expect that the dominant frequency of oscillations corresponds to the f_1 eigenfrequency ($\overline{\omega}_1 = \bar{a}_1^2$). On the contrary, for variants 2–6, the amplitude of the forced oscillations is determined by the $S_2(\bar{t})X_2(\bar{x})$ term and the dominant frequency of oscillations corresponds to the f_2 eigenfrequency ($\overline{\omega}_2 = \bar{a}_2^2$).

Figure 14 shows the function $v_1(\bar{t}, L) = L\bar{v}_1(\bar{t}, 1)$ for the forced oscillations of the free tail boom end (data are presented in mm) for different values of the parameter γ except of the resonance variant 4).

In general, one can note that the interaction between eigen and forced vibrations leads to two oscillation types having short ($\Delta\overline{T}_s$) and long ($\Delta\overline{T}_l$) dimensionless periods. Thus, all considered variants can be divided into two cases. Case I includes variants 1 and 6, for which $2N\overline{\omega} \neq \overline{\omega}_2$, and case II includes variants 2–5, for which $2N\overline{\omega} \approx \overline{\omega}_2$.

For case I, the short period $\Delta\overline{T}_s$ of oscillations is determined by the eigenfrequency f_1 ($\overline{\omega}_1 = 3.515625$) and oscillations of the free tail boom end have approximately constant amplitudes of about 0.1 mm (peak to peak relative deviation is approximately 50 microstrains). In general, this value of the oscillation amplitudes corresponds to estimations of paper [11] for static deformations of a light helicopter tail boom.

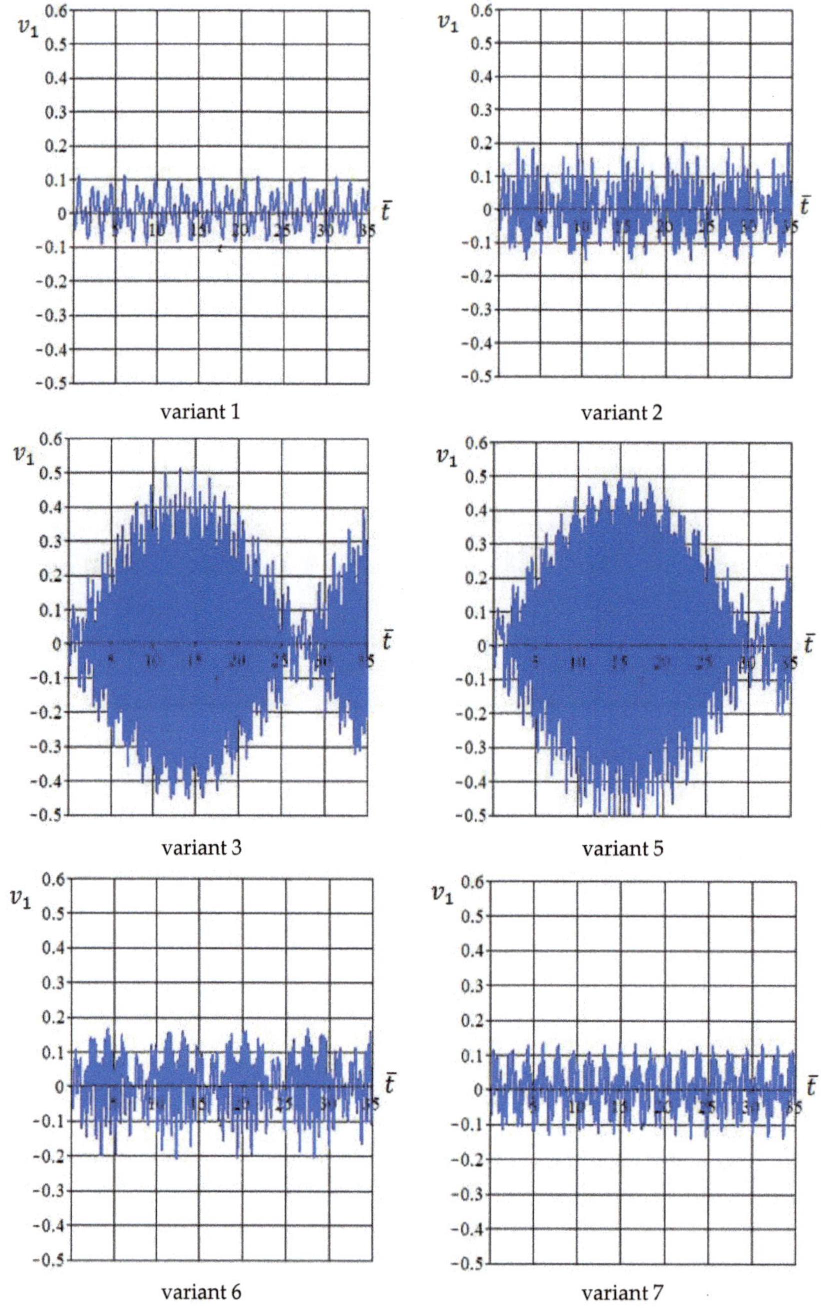

Figure 14. Forced free end tail boom oscillations $v_1(0, L)$ (in mm) at different γ values (variant numbers).

For all variants of case II, the short period $\Delta\overline{T}_s$ of the oscillations is determined by the eigenfrequency f_2 ($\overline{\omega}_2 = 22.033636$) and the long (interactional) oscillations period $\Delta\overline{T}_l$ depends on the γ parameter value. For variant 2, the period of interactional oscillations is approximately equal to $\Delta\overline{T}_l \approx 6$ (0.417 s) and $\Delta\overline{T}_l \approx 8$ (0.603 s) for variant 6. Maximum amplitude of the interactional oscillations is about 0.2 mm (peak to peak deviation is about 100 microstrains).

The period and amplitude of interactional oscillations increased as the value of $2N\overline{\omega}$ approaches the $\overline{\omega}_2$ value. For variant 3, the dimensionless period of interactional oscillations is approximately equal to $\Delta\overline{T}_l \approx 28$ (2.016 s), and for variant 5 takes place $\Delta\overline{T}_l \approx 32$ (2.35 s). The maximum amplitude of the interactional oscillations for variants 3 and 5 is about 0.5 mm (peak to peak deviation is approximately 250 microstrains).

6. Conclusions

The effect of the unsteady aerodynamics on the forced vibration and deformation of a helicopter tail-boom was considered. CFD modeling was used to compute the unsteady flow around the main rotor-fuselage, and then the aerodynamic loads were used in conjunction with the analytical structural model, based on the Euler-Bernoulli equation with one spatial coordinate. A solution of the Euler-Bernoulli was presented as a series of spatial and time coordinates, including four harmonics. The normal force coefficient acting on the tail boom surface was approximately determined based on CFD results for the rotor-fuselage interaction case. The aerodynamics were obtained assuming rigid tail boom.

The results of simulations showed that the amplitudes of the forced oscillations due to the main rotor blade rotation are relatively small. However, for a long operational flight period, deformations can become essential from the point of view of initiation of metal cracks, and delamination of structural beam elements made of composite materials.

For certain geometric parameters of the tail boom, the results demonstrate a resonance effect if the second blade passing harmonic is equal to the second tail boom eigenfrequency.

The tail boom was modeled as a thin-walled cylindrical structure with continuously distributed mass (no concentrated mass points). In the future, the proposed vibration model will be generalized for cylindrical structures with concentrated masses along their length.

Acknowledgments: The support of this work via the "State Tasks of the Education Ministry of Russia" grant (No. 9.1577.2017/PCH) is gratefully acknowledged.

Author Contributions: Alexander Kusyumov and George N. Barakos conceived the idea, designed and wrote the paper. Andrey Batrakov conducted CFD modeling, Sergey Kusyumov calculated tail boom deformations. Sergey Mikhailov co-designed and revised the paper.

Conflicts of Interest: The authors declare no conflict of interests.

Nomenclature

T = rotor thrust
D = fuselage drag
Cp = pressure coefficient
C_T = rotor thrust coefficient
C_D = fuselage drag coefficient
M_{tip} = tip Mach number
N = number of blades
q_∞ = free stream dynamic pressure
q_{tip} = blade tip dynamic pressure
R = rotor radius
$\overline{r}$ = normalized rotor radius
x = longitudinal tail boom coordinate
t = time coordinate
S_F = reference fuselage area
L = tail boom length
c_{Bz} = vertical load coefficient
E = Young's modulus
I = moment of inertia
m_L = mass per unit length
F_L = per unit length tail boom force
c_F = normal force coefficient
v = vertical deformation
v_0 = natural deformation
v_1 = forced deformation

Greek symbols

ρ_b = beam material density
ψ = rotor azimuth angle
ω = angular velocity of rotor

References

1. Crouse, L.; Leishman, J.; Bi, N. Theoretical and Experimental Study of Unsteady Rotor/Body Aerodynamic Interactions. *J. Am. Helicopter Soc.* **1990**, *37*, 55–65. [CrossRef]
2. Sydney, A.; Leishman, J.G. Measurements of Rotor/Airframe Interactions in Ground Effect under Simulated Brownout Conditions. In Proceedings of the American Helicopter Society 69th Annual Forum, Phoenix, AZ, USA, 21–23 May 2013.
3. Meerwijk, L.; Brouwer, W. Real-Time Helicopter Simulation Using the Blade Element Method. In Proceedings of the 17 European Rotorcraft Forum, Berlin, Germany, 24–27 September 1991.
4. Clough, R.W.; Penzien, J. *Dynamics of Structures*, 5rd ed.; Computers & Structures, Inc.: Berkeley, CA, USA, 2003; pp. 365–424. ISBN 978-0070113923.
5. Meirovitch, L. *Fundamentals of Vibrations*; McGraw-Hill International Edition: New York, NY, USA, 2001; pp. 374–458, ISBN 0-07-041345-2.
6. Weaver, W.; Timoshenko, S.P.; Young, D.H. *Vibration Problems in Engineering*, 5th ed.; John Wiley & Sons, Inc.: New York, NY, USA, 1990; pp. 422–433, ISBN 0-471-632287.
7. Rao, S.S. *Mechanical Vibrations*, 5rd ed.; Addison-Wesley Publishing Company: Boston, MA, USA, 1995; pp. 721–739, ISBN 978-0-13-212819-3.
8. Hsu, J.-C.; Lai, H.-Y.; Chen, C.K. Free Vibration of Non-Uniform Euler-Bernoulli Beams with General Elastically End Constraints Using a Domain Modified Decomposition Method. *J. Sound Vib.* **2008**, *318*, 965–981. [CrossRef]
9. Coskun, S.B.; Atay, M.T.; Ozturk, B. Transverse Vibration Analysis of Euler-Bernoulli Beams Using Analytical Approximate Techniques. In *Advances in Vibration Analysis Research*; Ebrahimi, F., Ed.; InTech: Vienna, Austria, 2011; pp. 1–22, ISBN 978-953-307-209-8.
10. Staley, J.A.; Sciarra, J.J. Coupled Rotor/Airframe Vibration Prediction Methods. In Proceedings of the Specialists Meeting on Rotorcraft Dynamics, Moffet Field, CA, USA, 13–15 February 1974; NASA Ames Res. Center Rotorcraft Dyn.: Moffett Field, CA, USA, 1974; pp. 81–90.
11. Rose, J.B.R.; Vetrivel, S. Structural Design and Analysis of Cost Effective Rotorcraft for Recovery Purposes. *Int. J. Eng. Trends Technol. (IJETT)* **2014**, *10*, 225–229. [CrossRef]
12. Krylov, A.N. *Vibrationships*, Shipbuilding Literature Edition; ONTINKTP: Leningrad-Moscow, USSR, 1936; pp. 326–365. (In Russian)
13. Steijl, R.; Barakos, G.; Badcock, K. A Framework for CFD Analysis of Helicopter Rotors in Hover and Forward Flight. *Int. J. Numer. Methods Fluids.* **2006**, *51*, 819–847. [CrossRef]
14. Batrakov, A.; Garipova, L.; Kusyumov, A.; Mikhailov, S.; Barakos, G. CFD Computational Fluid Dynamics Modeling of Helicopter Fuselage Drag. *J. Aircraft* **2015**, *52*, 1634–1643. [CrossRef]
15. Dehaeze, F.; Baverstock, K.D.; Barakos, G.N. CFD simulation of flapped rotors. *Aeronaut. J.* **2015**, *119*, 1561–1583. [CrossRef]
16. Woodgate, M.A.; Pastrikakis, V.A.; Barakos, G.N. Method for calculating rotors with active gurney flaps. *J. Aircraft* **2016**, *53*, 605–626. [CrossRef]
17. Carrion, M.; Woodgate, M.; Steijl, R.; Barakos, G. Implementation of All-Mach Roe-type Schemes in Fully Implicit CFD Solvers—Demonstration for Wind Turbine Flows. *Int. J. Numer. Methods Fluids* **2013**, *73*, 693–728. [CrossRef]
18. Lawson, S.; Barakos, G.N. Review of Numerical Simulations for High-Speed, Turbulent Cavity Flows. *Prog. Aerosp. Sci.* **2011**, *47*, 186–216. [CrossRef]
19. Kusyumov, A.; Mikhailov, S.; Garipov, A.; Nikolaev, E.; Barakos, G. CFD Simulation of Fuselage Aerodynamics of the "ANSAT" Helicopter Protype. *Trans. Control Mech. Syst.* **2012**, *1*, 318–324.
20. Batrakov, A.S.; Kusyumov, A.N.; Mikhailov, S.A.; Pakhov, V.V. A Study in Helicopter Fuselage Drag. In Proceedings of the 39th European Rotorcraft Forum, Moscow, Russia, 3–6 September 2013.
21. Wilcox, D.C. Re-assessment of the Scale-Determining Equation for Advanced Turbulence Models. *AIAA J.* **1988**, *26*, 1299–1310. [CrossRef]
22. Garipova, L.I.; Batrakov, A.S.; Kusyumov, A.N.; Mikhailov, S.A.; Barakos, G. Estimates of Hover Aerodynamics Performance of Rotor Model. *Russ. Aeronaut.* **2014**, *57*, 234–244. [CrossRef]

23. Caradonna, F.X.; Tung, C. Experimental and Analytical Studies of a Model Helicopter Rotor in Hover. In Proceedings of the 6th European Rotorcraft and Powered Lift Aircraft Forum, Bristol, UK, 16–19 September 1980.
24. Batrakov, A.S.; Kusyumov, A.N.; Barakos, G. Simulation of Flow around Fuselage of Helicopter Using Actuator Disc Theory. In Proceedings of the 29 ICAS Conference, St. Petersburg, Russia, 7–12 September 2014.

© 2017 by the authors. Licensee MDPI, Basel, Switzerland. This article is an open access article distributed under the terms and conditions of the Creative Commons Attribution (CC BY) license (http://creativecommons.org/licenses/by/4.0/).

applied sciences

MDPI

Article

Experimental Tests and Aeroacoustic Simulations of the Control of Cavity Tone by Plasma Actuators

Hiroshi Yokoyama *, Isamu Tanimoto and Akiyoshi Iida

Department of Mechanical Engineering, Toyohashi University of Technology, Toyohashi 441-8580, Japan; tanimoto@aero.me.tut.ac.jp (I.T.); iida@me.tut.ac.jp (A.I.)

* Correspondence: h-yokoyama@me.tut.ac.jp; Tel.: +81-532-44-6665

Received: 18 July 2017; Accepted: 3 August 2017; Published: 4 August 2017

Featured Application: Noise reduction in high-speed transport vehicles.

Abstract: A plasma actuator comprising a dielectric layer sandwiched between upper and lower electrodes can induce a flow from the upper to lower electrode by means of an externally-applied electric field. Our objective is to clarify the mechanism by which such actuators can control the cavity tone. Plasma actuators, with the electrodes elongated in the streamwise direction and aligned in the spanwise direction, were placed in the incoming boundary of a deep cavity with a depth-to-length ratio of 2.5. By using this experimental arrangement, the amount of sound reduction ("control effect") produced by actuators of differing dimensions was measured. Direct aeroacoustic simulations were performed for controlling the cavity tone by using these actuators, where the distributions of the body forces applied by the actuators were determined from measurements of the plasma luminescence. The predicted control effects on the flow and sound fields were found to agree well with the experimental results. The simulations show that longitudinal streamwise vortices are introduced in the incoming boundary by the actuators, and the vortices form rib structures in the cavity flow. These vortices distort and weaken the two-dimensional vortices responsible for producing the cavity tone, causing the tonal sound to be reduced.

Keywords: cavity tone; direct aeroacoustic simulation; aeroacoustics; plasma actuators; flow control; noise control; vortices; wind tunnel experiments; acoustic sources

1. Introduction

Self-sustained oscillations in a flow over a cavity as shown in Figure 1—such as a sunroof of an automobile or various gaps between parts of a high-speed transportation vehicle—can radiate intense tonal noise as a cavity tone. This high-intensity noise also occurs at a single peak frequency, making it a very unpleasant for many people. It is, thus, important to develop methods to suppress such cavity tones.

Many researchers over the past 50 years have investigated the mechanism of acoustic radiation from self-sustained oscillations in cavity flows. Rossiter [1] described an oscillation mechanism similar to that presented for edge tones by Powell [2]. In this mechanism, the interactions of vortices with the downstream edge of the cavity radiate acoustic waves, which cause the formation of new vortices at the upstream edge. Furthermore, an acoustic resonance sometimes occurs in the cavity—such as a one-quarter-wavelength-depth mode—making the cavity tone more intense [3].

Zhuang et al. [4] investigated the control of the cavity tone by blowing jets at the upstream edge of the cavity. Yokoyama et al. [5] showed that the cavity tone can be reduced by blowing jets aligned across the span in the upstream boundary layer, and the amount of reduction was affected by the spacing of the jets.

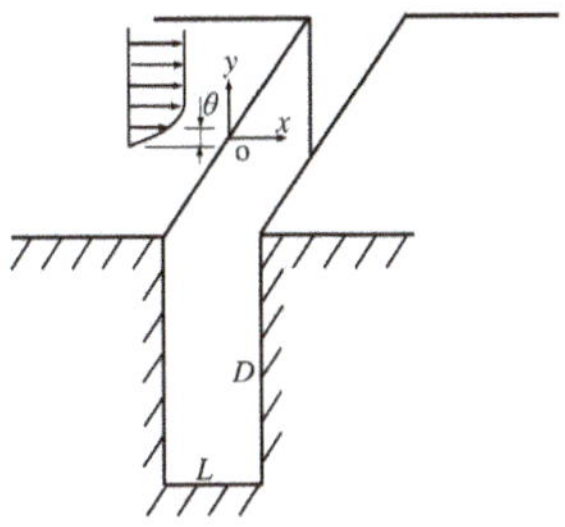

Figure 1. Configuration of the flow over a cavity.

Control of the cavity tone by a plasma actuator (PA) has also been investigated [6,7]. Figure 2 shows a typical configuration of a dielectric-barrier-discharge plasma actuator. It comprises an upper exposed and a lower embedded electrode, with a dielectric layer sandwiched between the electrodes. By applying a high AC voltage to the electrodes, momentum is transferred from the charged plasma particles to the neutral gas through collisions between the plasma components and neutral air molecules. As a result, a flow is induced from the upper electrode toward the lower one. Plasma actuators have many advantages: they are simple configurations without moving parts, have fast time response and low mass, with the ability to be attached easily onto objects, as compared with other flow-control actuators, such as a piezo-actuator, an impinging jet, or a synthetic jet.

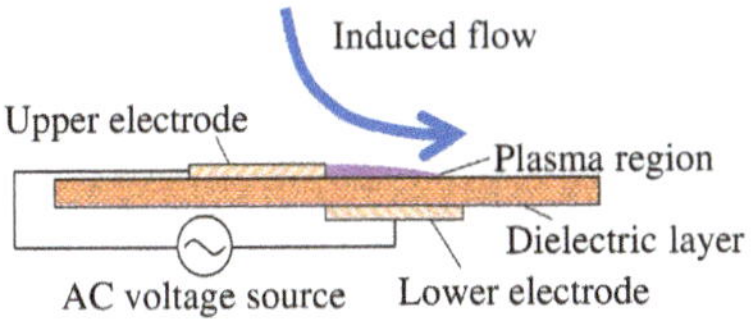

Figure 2. Typical configuration of a dielectric-barrier-discharge plasma actuator.

Huang et al. [6] succeeded in reducing the cavity tone by using spanwise-aligned plasma actuators, with the electrodes elongated in the streamwise direction ("streamwise plasma actuators"), which introduced a spanwise variation of velocity in the cavity flow. However, the sound-reduction mechanism of this approach has not been clarified in detail.

The objective of the present investigation is to clarify the mechanism by which streamwise plasma actuators reduce the cavity tone. In preliminary experiments, the effects of the spanwise width and pitch of the lower electrode on the control of the cavity tone were studied. In Section 2, both our experimental methodologies and the simulations performed to clarify the mechanism through which the plasma actuators control the intensity of the cavity tone are discussed. The comparison of predicted results with those measured is discussed in Section 3. The predicted and measured results are discussed in detail in Section 4. In Section 5, conclusions are summarized.

2. Methodologies

2.1. Experimental Tests

2.1.1. Wind Tunnel and Test Section

Figure 3 shows the experimental setup for our tests. The experiments were carried out using a suction-type, low-noise wind tunnel with a rectangular test section having a cross-section with dimensions of 150 mm × 75 mm. The intensity of freestream turbulence was less than 0.5%, and the background-noise level was 55.2 dB (A) at the freestream velocity of 30 m/s.

The origin of coordinates was located at the mid-span of the upstream edge of the cavity. The x, y, and z axes were oriented in the streamwise, normal, and spanwise directions, respectively.

The cavity length was L = 20 mm and the depth-to-length ratio was D/L = 2.5. From such small gaps between parts of an actual high-speed transportation mode, such as an automobile, tonal sound often radiates as cavity tone. The test section of the cavity was terminated in the spanwise direction by end walls composed of porous plates, which suppressed acoustic resonances in the spanwise direction. The width of the test section was W/L = 7.5.

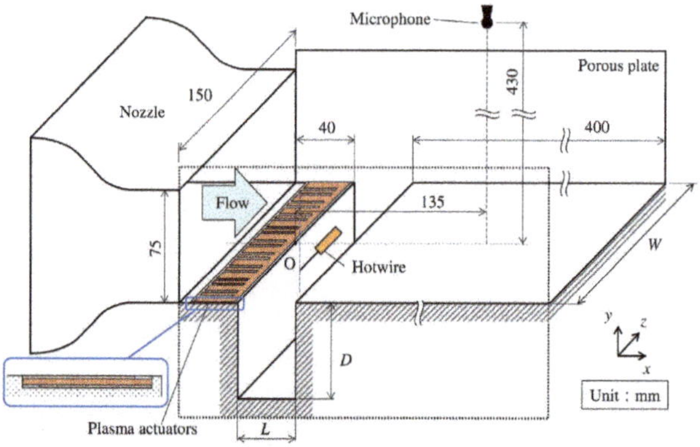

Figure 3. Experimental setup for the measurement of control effects on cavity flow and tone.

2.1.2. Measurement of Sound and Flow Fields

The magnitude of the acoustic pressure in the far field (x/L = 6.75, y/L = 21.5) was measured with a non-directional $\frac{1}{2}$-inch microphone (UC-53A, RION, Tokyo, Japan) and a precision sound-level meter (NL-52, RION, Tokyo, Japan) for a freestream flow velocity U_0 = 10–45 m/s. The measurement time required for data acquisition was 30 s, and the sampling frequency was 40 kHz.

The flow induced by the plasma actuators was also visualized by using a fog generator (Safex Fog Generator 2010, Dantec Dynamics A/S, Skovlunde, Denmark), with tracer particles (D/SPEZIAL, Safex, Tangstedt, Germany), in the absence of a freestream flow. The generated fog was illuminated by a green laser sheet, and photographs of the visualized flow were taken with a digital camera (Nikon D5200, Nikon, Tokyo, Japan).

2.2. *Test Case Descriptions*

In the experiments, the control effects of the actuators on the sound pressure were clarified while changing the freestream velocities, U_0 = 10–45 m/s, while the simulations were performed at U_0 = 30 m/s. The actuators are shown in Figure 3. The Reynolds number based on the cavity length was $\mathrm{Re} \equiv U_0 L/\nu = 1.3 \times 10^4$–$6.0 \times 10^4$, and the Mach number was $M \equiv U_0/a$ = 0.029–0.131, where ν and a are the dynamic viscosity and the speed of sound in air, respectively. In addition, to clarify the nature of the flow induced by the actuators, flow visualizations and simulations were performed in the absence of a freestream flow.

Figure 4 shows the measured sound-pressure spectra for freestream velocities U_0 = 15–45 m/s at the location x/L = 6.75 and y/L = 21.5 without control. The frequency resolution of the spectral analysis was 4.88 Hz. As shown in this figure, the tonal sound at the fundamental frequency f = 1500 Hz becomes most intense at a velocity of 30 m/s, where the existence of an acoustic resonance was confirmed by the phase distributions of the pressure in the cavity [8].

In both our experiments and simulations, the incoming boundary layer was laminar, and the velocity profile was confirmed to agree with the Blasius profile. The momentum thickness was

$\theta/L = 0.0071$ for a freestream velocity of $U_0 = 30$ m/s at the upstream edge of the cavity, where this value was measured and found to be consistent with that predicted for flow over a flat plate without a cavity.

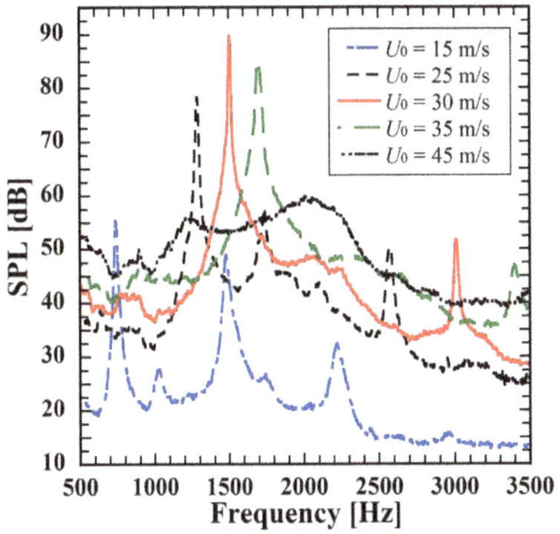

Figure 4. Measured sound-pressure spectra without control at U_0 = 15, 25, 30, 35, 45 m/s. SPL is the sound-pressure level in decibels

2.3. Plasma Actuator Control

2.3.1. Control by Streamwise Plasma Actuators

As shown in Figure 5, the plasma actuators consist of lower and upper copper electrodes 18 μm thick, with a dielectric layer of polyimide 200 μm thick sandwiched between them. The lower electrodes were placed in the spanwise direction between the upper electrodes. The actuators were designed and built by the authors. As discussed in Section 3.1, these actuators induce pairs of streamwise vortices with spanwise widths equal to those of the lower electrodes. The electrodes were arranged symmetrically in the spanwise direction with respect to the mid-span ($z = 0$).

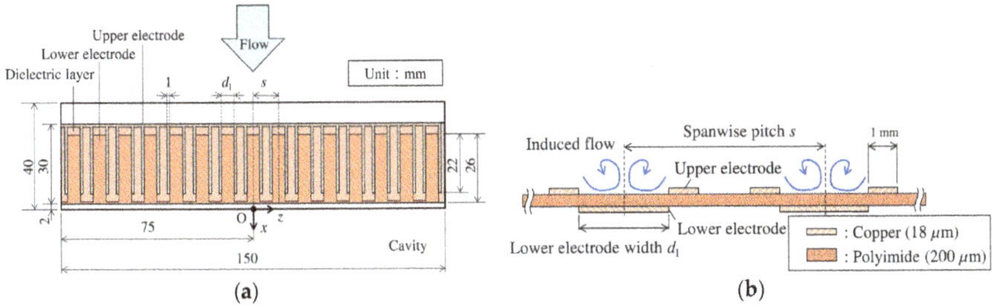

Figure 5. Configuration of the streamwise plasma actuators: (**a**) top view; and (**b**) view in a cross-section perpendicular to the streamwise direction.

An AC voltage at the frequency f_{pa} = 4.2 kHz was applied to the electrodes. To clarify the effects of the applied voltage on the control, it was changed in the range from E_{pa} = 4 $kV_{p\text{-}p}$ to 5.5 $kV_{p\text{-}p}$ in the experiments. Addtionally, the effects of the pitch of the actuators, s = 4, 10 mm, and the spanwise width of the lower electrode, d_l = 1, 3, 5, 7 mm, on the control were investigated. Moreover, to clarify the nature of the control on the cavity flow and tone, aeroacoustic simulations were performed, as well as flow measurements, both with and without control. An actuator with E_{pa} = 5 kV$_{p\text{-}p}$, s = 10 mm (s/L = 0.5, s/θ = 71), and d_l = 3 mm was utilized at U_0 = 30 m/s, where substantial reduction of the intense sound was achieved.

2.3.2. Flush-Mounted Plasma Actuators

To minimize the influence on the flow and on noise generation of the protrusion of the actuators from the wall, a flush-mounted actuator, with the dielectric layer embedded in the model, was utilized so that the protrusion was only the 18 μm thickness of the upper electrode. In order to maintain the sharpness of the cavity edge, the actuator was set back from the upstream edge by a space of 0.1L in the upstream direction.

Figure 6 compares the sound-pressure levels, at the fundamental frequency, of the cavity flow both with and without the actuator; the difference was negligibly small. Consequently, this flush-mounted actuator does not affect the evaluation of control effects on the cavity tone obtained using the low-noise wind tunnel.

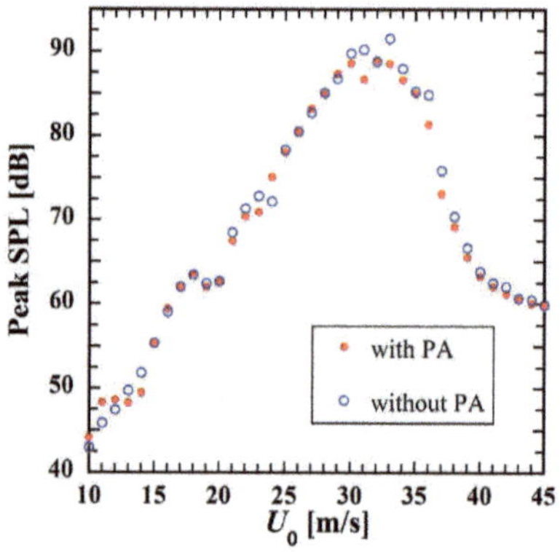

Figure 6. Comparison of the sound-pressure level SPL of the cavity flow both with and without a flush-mounted actuator.

2.4. Numerical Simulations

2.4.1. Governing Equations and Finite-Difference Schemes

To clarify the fluid-acoustic interactions in the cavity flows, the flow and acoustic fields were simultaneously simulated by solving the three-dimensional compressible Navier-Stokes equations with mass and energy conservation laws in conservation form:

$$\mathbf{Q}_t + (\mathbf{E} - \mathbf{E}_\mathrm{v})_x + (\mathbf{F} - \mathbf{F}_\mathrm{v})_y + (\mathbf{G} - \mathbf{G}_\mathrm{v})_z = \mathbf{D}, \tag{1}$$

$$\mathbf{D} = \begin{pmatrix} 0 \\ s_x(x,y,z,t) \\ s_y(x,y,z,t) \\ s_z(x,y,z,t) \\ D_c(s_x u + s_y v + s_z w) \end{pmatrix}, \tag{2}$$

where Q is the vector of the conserved variables, E, F, G are the inviscid flux vectors, and E_v, F_v, G_v are the viscous flux vectors. To reproduce the body force produced by the actuator, the term D representing the body force and the power added by it to the unit volume are included on the right-hand side of the equation [9]. Details of the body forces are described in next subsection.

The spatial derivatives were evaluated using a sixth-order-accurate, compact, finite-difference scheme (fourth-order accurate on the boundaries) [10]. Time integration was performed with a third-order-accurate Runge-Kutta method.

The flow in the cavity became turbulent under the given conditions, although the incoming boundary layer was laminar. To reduce the computational cost, large-eddy simulations (LES) were performed in the present study. No explicit subgrid-scale (SGS) model was used. The turbulent energy at the grid-scale (GS) that should be transferred to SGS eddies was dissipated by a 10th-order spatial filter described below. A number of studies [11–13] have shown that this approach, which combines low-dissipation discretization schemes with filtering, correctly reproduces turbulent flows. This filter, which also removes numerical instabilities, is given by [14]:

$$\alpha_f \hat{\psi}_{i-1} + \hat{\psi}_i + \alpha_f \hat{\psi}_{i+1} = \sum_{n=0}^{5} \frac{a_n}{2} (\psi_{i+n} + \psi_{i-n}), \tag{3}$$

where ψ is a conserved quantity, and $\hat{\psi}$ is the filtered quantity. For the coefficients a_n, the same values as Gaitonde and Visbal [15] were utilized, and the value of the parameter α_f was set to be 0.45. Cavity flows with incoming laminar and turbulent boundary layers can be predicted with high accuracy using the methods described above [16].

2.4.2. Body Force Exerted by Plasma Actuators

The body force and the power exerted by the actuators were modeled by including the quantities s_x, s_y, s_z and $us_x + vs_y + ws_z$ in the Navier-Stokes equations. For the body-force vector, the expression as follows was used:

$$\begin{pmatrix} s_x \\ s_y \\ s_z \end{pmatrix} = A\rho_c(x,y,z) \begin{pmatrix} \varphi_x(x,y,z) \\ \varphi_y(x,y,z) \\ \varphi_z(x,y,z) \end{pmatrix} \sin^2(2\pi f_{\mathrm{pa}} t), \tag{4}$$

$$A = 1 \; (\sin(2\pi f_{\mathrm{pa}} t) \leq 0), \; -0.4 (\sin(2\pi f_{\mathrm{pa}} t) > 0), \tag{5}$$

where ρ_c is the plasma density, and $\boldsymbol{\varphi}$ is the electric field. The square of the sine function in Equation (4) means that the plasma discharge occurs twice in one period, as in the body-force model of [9].

The distributions of plasma density and electric field were computed using the method proposed by Suzen [17]. In computing the plasma density, the Debye length was taken to be 0.7 mm, and the distributions of plasma density between the exposed upper electrodes were determined from the luminescence of the plasma.

Figure 7 shows the distributions of plasma density determined from the luminescence measurements along with those predicted near the wall (y = 0.05 mm), where the horizontal direction is the spanwise position from the midpoint of the embedded electrode (z = z_{pa}), which is non-dimensionalized using the pitch of the actuators. As shown in the figure, the predicted distributions agree well with those determined in the experiments.

The coefficient A in Equation (4) was introduced to represent the change of the amount and direction body forces between positive and negative applied voltages. The value of A was set to be the same as that in [18], where simulations by the above-described method were employed to reproduce the induced flow in an experiment by Forte [19].

The simulated distributions of the body forces are shown in Figure 8. The most intense body force is generated around the edges of the exposed electrodes. The non-dimensional maximum body force was taken to be $D_c = \rho_{c,\max} E_{\mathrm{pa}} / (\rho_0 U_0{}^2 / L) = 2.8 \times 10^{-6}$ so that control effects on the mean velocity profile in the cavity were in good agreement with those measured in the experiments discussed in Section 3.1.

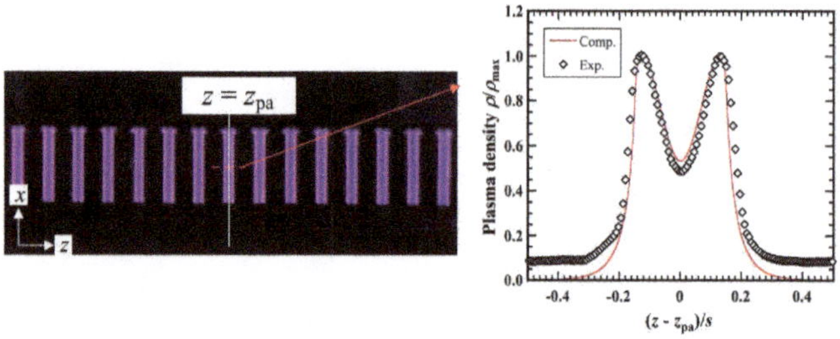

Figure 7. Plasma distributions. Photograph of the plasma luminescence (**left**). Predicted distributions of plasma density near the wall (y = 0.05 mm), compared with those determined from the plasma luminescence in the photograph (**right**).

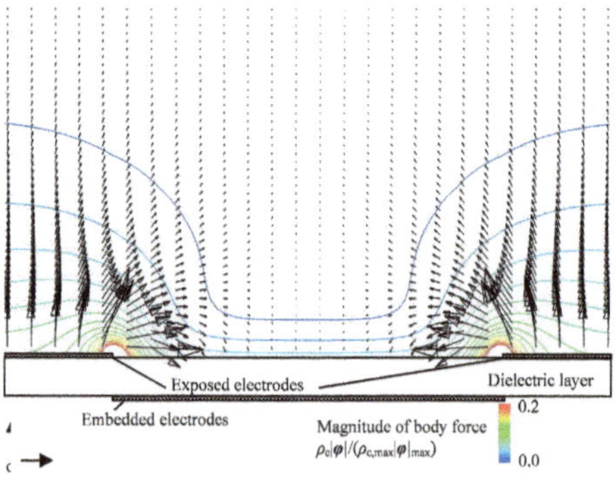

Figure 8. Predicted vectors of the body force and contours of the magnitude of the force vector.

2.4.3. Computational Grid

Figure 9 shows the computational grid for the flow over the cavity. The spanwise length $W_c/L = 3.0$ of the computational domain was sufficiently wide to reproduce the effects of control by an actuator with pitch $s/L = 0.5$, with six sets of electrodes included in the computation. The spanwise grid resolution was $\Delta z/L = 1/80$. The streamwise resolution in the cavity was $\Delta x/L = 1/100$, and the normal resolution near the wall in the incoming boundary layer and in the shear layer in the cavity was $\Delta y_{min}/L = 1/400$. The grid was stretched into the far field in both the normal and streamwise directions. The grid resolution in the x-y cross-section in the cavity was the same as that in our past research [8], where the predicted flow and sound fields were found to be in good agreement with those measured. The total number of grid points was 55 million.

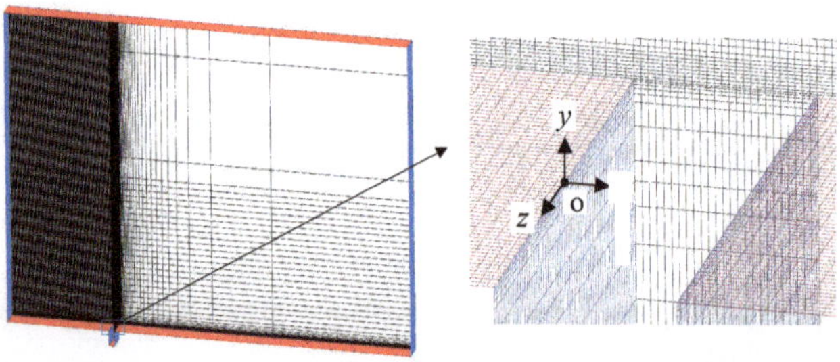

Figure 9. Computational grid, where every fourth grid line is shown for clarity.

2.4.4. Boundary Conditions

Figure 10 shows the boundary conditions. Non-reflecting boundaries [20–22] at the inflow and outflow boundaries were utilized so that acoustic waves pass the boundaries smoothly. On the lower boundary at the upstream side of the cavity, the boundary condition was changed from a slip wall to a non-slip and adiabatic wall. The position of this change was determined so as to make the thickness of the incoming boundary layer equal to that measured. Periodic boundary conditions were used in the spanwise direction. A uniform steady flow was asymptotically imposed in the inflow buffer region ($x/L \leq -12.5$).

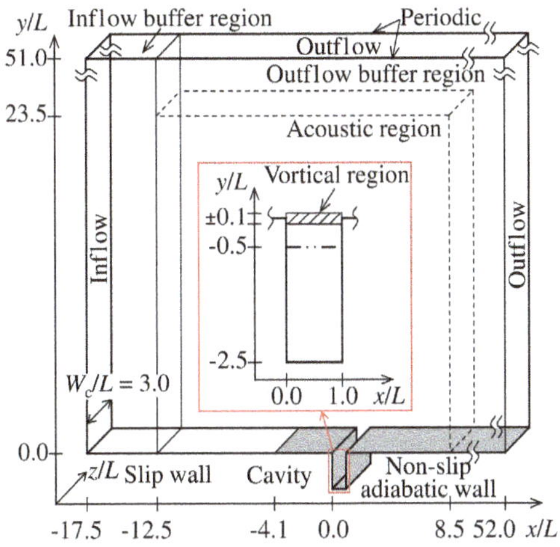

Figure 10. Computational domain and boundary conditions.

2.4.5. Prediction of the Far Acoustic Field

The porous Ffowcs, Williams, and Hawkings (FW-H) method [23–25] was used to predict the acoustic pressure at the measurement point ($x/L = 6.75$ and $y/L = 21.5$). The pressure p was sampled at a semi-cylindrical surface of radius $r = 2.5L$, near the center of the cavity ($x/L = 0.5$ and $y = 0$).

A spanwise computational domain of width $W_c = 3L$, which was chosen to be smaller than that of the experiment, $W = 7.5L$, was utilized in order to reduce computational resources. To take the effects of this difference into account, the sound-pressure levels, SPL (f), were corrected by using the equivalent coherence length, L_c (f), which was computed based on the coherence of the velocity along the center of the cavity ($x/L = 0.5$, $y = 0$) in the spanwise direction. The detailed method is described in [5].

3. Comparison and Validation

3.1. Flow Fields

Figure 11a shows a visualization of the induced flow in the y-z cross-section produced by activating the actuators in the absence of a freestream flow. As shown in the figure, a pair of vortices with a streamwise (x) axis are induced by each actuator, composed of two upper electrodes and one lower electrode. Figure 11b shows the predicted velocity vectors color-coded by the vertical velocity component under the same conditions. The streamwise vortices measured in the experiments also occur in the present simulation. The predicted maximum induced velocity was roughly 2 m/s.

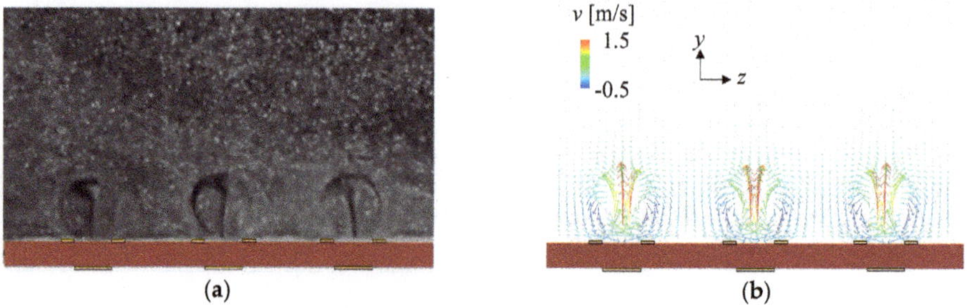

Figure 11. Vortices induced by actuators in the absence of a freestream flow: (**a**) visualized flow in the experiment; and (**b**) the predicted velocity vectors color-coded by values of the vertical velocity.

The predicted mean velocity profiles at the position $x/L = 0.05$ and $z = z_{pa}$ for $U_0 = 30$ m/s, both with and without control are compared with those measured. The time-series velocity, measured by the hot-wire anemometer in our experiment, was given by $u_h = (u^2 + [0.5v]^2)^{0.5}$ in the computation, in the same way as in our previous research [26], and the time-averaged values, U_h, were computed. Figure 12 shows that the predicted profile agrees well with that measured for cases both with and without control. The free shear layer is moved in the upward direction by the control. This is because the incoming boundary is moved upward due to the streamwise vortices induced by the control.

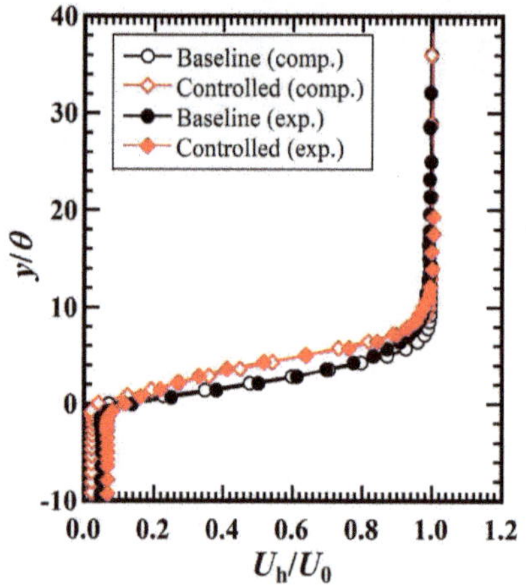

Figure 12. Predicted and measured mean velocity profiles with and without control ($x/L = 0.05$, $z = z_{pa}$).

3.2. Sound-Pressure Spectra

Figure 13 shows the measured and predicted sound-pressure spectra both with and without control, where the frequency resolution is $\Delta St = 0.1$ ($\Delta f = 156$ Hz), where St represents the Strouhal number based on the cavity length and the freestream velocity. The spectra were averaged 9374 and 18 times, respectively, in the experimental and computational results. The predicted fundamental frequency around St = 1.0 agrees well with that measured both with and without control. The sound-pressure level at the fundamental frequency also is in good agreement with that measured, both with and without control, where the level is reduced by the control. The predicted and measured reduction levels were 5 and 6 dB, respectively.

The discussions in the previous and present subsections confirm that the present computations adequately predict the flow and sound fields, both with and without control.

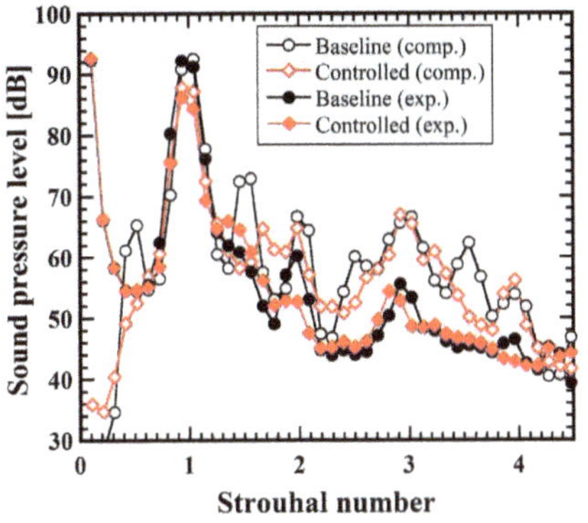

Figure 13. Predicted and measured sound-pressure spectra, both with and without control.

4. Results and Discussions

4.1. Measured Geometric Effects of Actuators on Control

The variation of the sound-reduction level at the fundamental frequency of 1500 Hz as a function of the voltage E_{pa} applied to the actuators was measured, where the spanwise pitch of the lower electrode was either s = 4 mm or s = 10 mm, and the fixed width of the lower electrode was d_l = 3 mm. Figure 14a shows the measured variation. The frequency resolution of spectral analysis was Δf = 4.88 Hz and it should be noted that the reduction level was different from that in the previous section due to the difference of the frequency resolution.

As shown in Figure 14a, the amount of noise reduction is greater for higher applied voltages, particularly in the range from E_{pa} = 4 $kV_{p\text{-}p}$ to 5.5 $kV_{p\text{-}p}$. The sound level drops sharply at lower voltages for the narrower pitch s = 4 mm, as compared to the case with s = 10 mm. For s = 4 mm, the sound reduction level of 31 dB was achieved at E_{pa} = 5 $kV_{p\text{-}p}$. However, for a fixed cavity width, more actuators and greater power consumption are necessary for narrower pitches. In the next subsection, the sound-reduction mechanism is discussed for s = 10 mm and E_{pa} = 5.0 $kV_{p\text{-}p}$.

According to the investigation by Yokoyama et al. [5], the cavity tone was also reduced by pairs of introduced streamwise vortices by blowing jets in the incoming boundary layer. Additionally, the results at the jet velocity of V_j/U_0 = 0.07, which roughly corresponds to the jet velocity (2 m/s) induced by the present plasma actuators, show that the control effects became smaller for finer spanwise displacement of the jets, such as s/L = 0.1 and 0.25. The maximum reduction level was achieved at s/L = 0.5. This was because the induced longitudinal vortices were weakened for the finer displacement.

Meanwhile, the present plasma actuators can most largely reduce cavity tone with a finer displacement of s/L = 0.2. This difference from the results of blowing jets is possibly because the electrodes of the actuators are elongated into the streamwise direction making it easier to introduce intense streamwise vortices, while the blowing jets were ejected from simple circular holes. Deliberately SETTING up the inlet shape and blowing direction for the jets for the introduction of the streamwise vortices possibly increases the effects.

The comparison of the present results with those of the control by blowing jets indicates that the present plasma actuators can effectively introduce longitudinal vortices and reduce cavity tone with a finer displacement. Meanwhile, for a fixed cavity width, more actuators and greater power consumption are necessary for a narrower displacement. The exact estimation of power consumption remains a problem for the future.

To clarify the effects of the lower-electrode width on control of the cavity tone, the sound-reduction levels were measured for lower-electrode widths of d_l = 1, 3, 5, and 7 mm, at a constant pitch of

s = 10 mm, as shown in Figure 14b. The results show that sound reduction was obtained and that the influence of the lower-electrode width on sound reduction was small, except for the narrowest width, d_l = 1 mm. Streamwise vortices are not induced at this narrow width, possibly because the body force is not sufficiently large, due to too narrow a width for expansion of the plasma between the exposed electrodes [27].

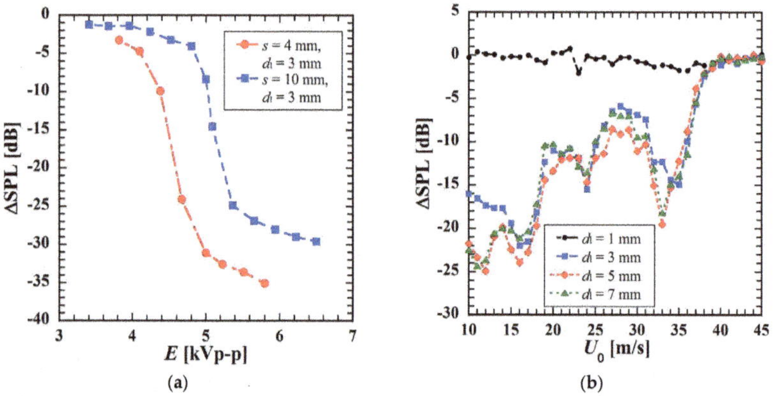

Figure 14. (**a**) The variation of the sound-reduction level, in decibels, at the fundamental frequency of 1500 Hz as a function of the applied voltage; and (**b**) the variation of the sound-reduction level for different lower-electrode widths as a function of the freestream velocity U_0.

4.2. Control Effects on Flow Structures and Pressure Fields

4.2.1. Vortical Structures

Figure 15 shows the predicted vortical structures with and without control as iso-surfaces of the second invariant $q \equiv ||\Omega||^2 - ||S||^2$, where Ω and S are the anti-symmetric parts and symmetric parts, respectively, of the velocity gradient tensor for the phase-averaged flow fields at the phase where the vortex collides with the downstream edge of the cavity. The iso-surfaces are color-coded by the values of the streamwise vorticity.

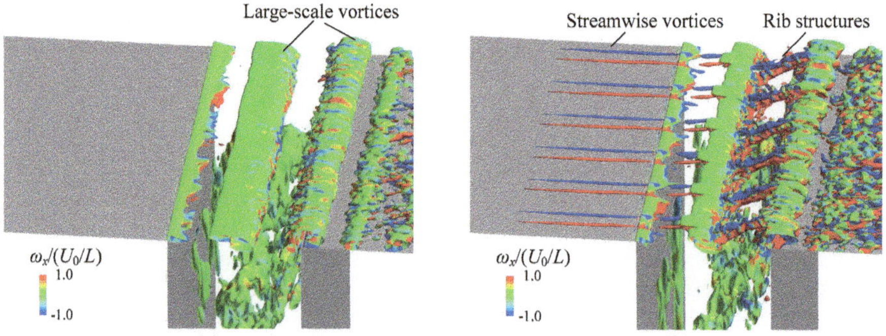

Figure 15. Iso-surfaces of the second invariant of $q/(U_0/L)^2$ = 0.45 color-coded by values of the streamwise vorticity for phase-averaged flow fields at U_0 = 30 m/s at the phase where the vortex collides with the downstream edge of the cavity: (**a**) baseline flow; and (**b**) controlled flow.

As shown in Figure 15a, two-dimensional large-scale vortices are generated in the baseline cavity flow. Low-pressure regions are generated in the large-scale vortices, and an expansion wave is radiated due to the collision of the vortices with the downstream edge of the cavity [16]. Figure 15b shows that streamwise vortices are induced by the actuators in the incoming boundary layer and form rib structures between the large-scale vortices.

Figure 16 shows the iso-surfaces of the component of the second invariant q_z, which was computed from Equation (6):

$$q_z \equiv |\omega_z|^2 - \left\| \begin{pmatrix} S_{xx} & S_{yx} \\ S_{xy} & S_{yy} \end{pmatrix} \right\|^2, q_z \equiv |\omega_z|^2 - \left\| \begin{pmatrix} S_{xx} & S_{yx} \\ S_{xy} & S_{yy} \end{pmatrix} \right\|^2, \quad (6)$$

where ω_z is the spanwise vorticity and S_{xx}, S_{xy}, and S_{yy} are the elements of the symmetric parts of the velocity gradient tensor. This quantity is related to vortices with spanwise axes and is color-coded by the spanwise vorticity in the figure. Figure 16 also shows the contours of streamwise vorticity in the shear layer (y/θ = 2.9).

As shown in Figure 16a, vortical structures with spanwise axes are apparent in the baseline flow. These vortices correspond approximately to the above-mentioned large-scale vortices. The periodic collision of these vortices with the downstream edge of the cavity produces the intense cavity tone. Figure 16b shows that these vortices are distorted by the streamwise vortices in the controlled flow.

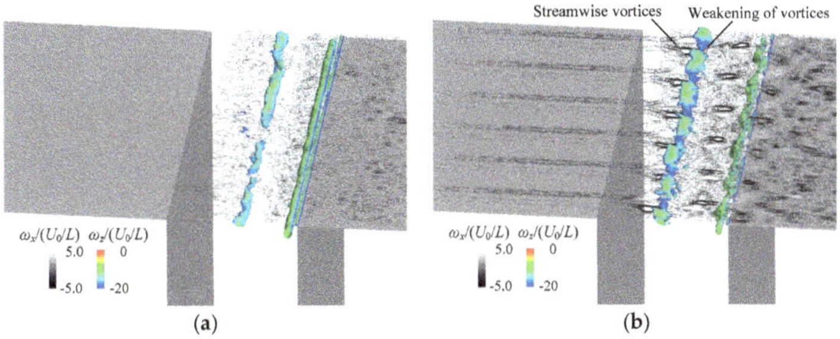

Figure 16. Iso-surfaces of the component of the second invariant related to vortices with a spanwise axis $q_z/(U_0/L)^2 = -20$ color-coded by the spanwise vorticity, together with contours of the streamwise vorticity in the shear layer (y/θ = 2.9) at the phase where the vortex collides with the downstream edge of the cavity: (**a**) baseline flow; and (**b**) controlled flow.

4.2.2. Power Spectra of Velocity Fluctuations

Figure 17 shows the predicted power spectra of velocity fluctuations at x/L = 0.5 and $z = z_{pa}$ at the peak of maximum power for y/θ = 6.4 and 3.6, respectively, with and without control, along with the measured data. The figure shows that the power at the fundamental frequency is weakened by the control in both computations and experiments. This corresponds to the weakening of the above-mentioned large-scale vortices with spanwise axes that produce the cavity tone.

4.2.3. Spanwise Coherence

To clarify further the nature of the vortical structures, the coherence between two points separated by spanwise distances Δz at x/L = 0.5 and y = 0 was examined. Figure 18 shows the spanwise coherence of the streamwise velocity with and without control. Although the coherence shows locally-high values, depending on the pitch s/L = 0.5 of the actuators, overall it is decreased.

Figure 19 shows the phase-averaged fluctuation pressure with and without control at the phase where the vortex collides with the downstream edge of the cavity. In the baseline flow, the distribution of the pressure is coherent in the spanwise direction. At the collision of the large-scale vortical structures with the downstream edge of the cavity, the radiated sound is reinforced in the spanwise direction by the coherent pressure fluctuations. In the controlled flow, however, the fluctuation pressure varies in the spanwise direction. This variation also contributes to the weakening of the cavity tone.

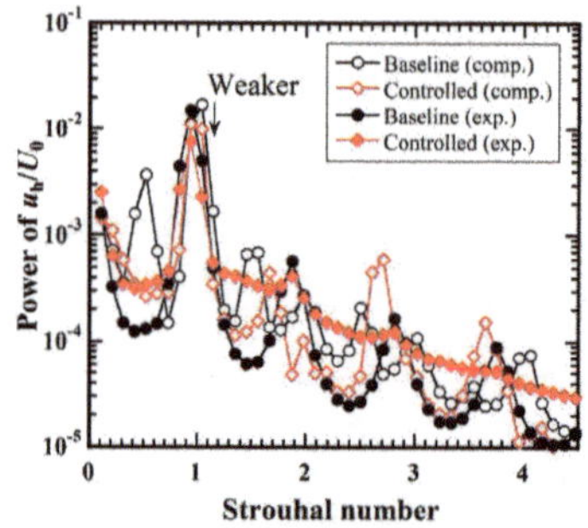

Figure 17. Predicted power spectra of velocity, u_h, with and without control at the peak of maximum power for y/θ = 6.4 and 3.6 at x/L = 0.5 and $z = z_{pa}$, with and without control, respectively. The measured spectra at the peak of maximum power for y/θ = 5.7 and 2.8 at the same streamwise and spanwise positions of x/L = 0.5 and $z = z_{pa}$ with and without control, respectively, are also shown.

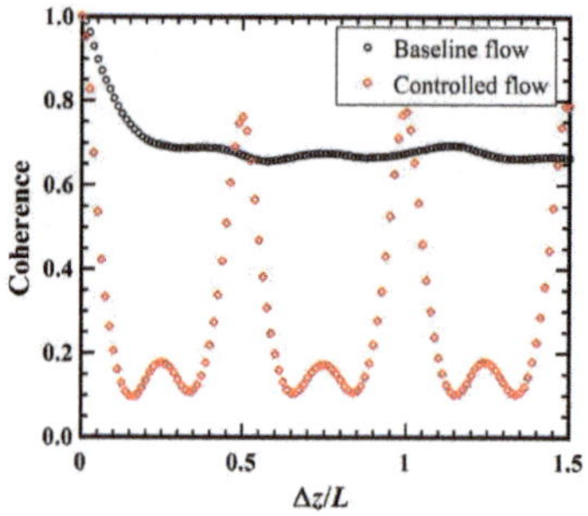

Figure 18. Spanwise coherence of streamwise velocity, u, at x/L = 0.5 and y = 0 for baseline and controlled flows.

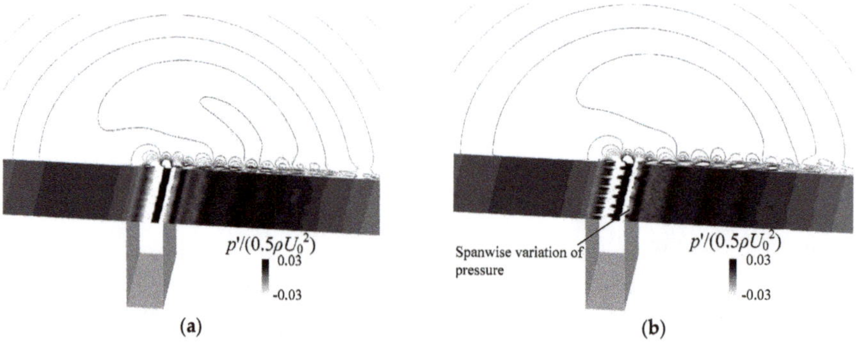

Figure 19. Contours of the fluctuation pressure at U_0 = 30 m/s for phase-averaged flow fields at the phase where the vortex collides with the downstream edge of the cavity: (**a**) baseline flow; and (**b**) controlled flow.

5. Conclusions

To clarify the mechanism by which streamwise plasma actuators aligned in the spanwise direction reduce the cavity tone with the acoustic resonances, direct aeroacoustic simulations were performed, as well as wind tunnel experiments. The predicted flow and acoustic fields, both with and without control, are in good agreement with those measured. To the authors' knowledge, this is the first time that direct aeroacoustic simulations of the control of aerodynamic noise by plasma actuators have been performed, including comparisons with experimental data.

On the measured effects of lower electrode width on the control, the cavity tone was effectively reduced for a lower-electrode width of d_l = 3, 5, and 7 mm, while the control effects on the cavity tone were negligible for d_l = 1 mm. The predicted and measured flow fields show that streamwise vortices are induced by the actuators for d_l = 3 mm in the absence of a freestream.

The results show that aerodynamic sound can be controlled at the relatively high freestream velocity of 30 m/s by 31 dB by plasma actuators with the spanwise fine pitch of actuators of s = 4 mm and d_l = 3 mm at the applied voltage of E_{pa} = 5 $kV_{p\text{-}p}$, even though they can only induce flows with a maximum velocity of 2 m/s. In comparison with the control by blowing jets, the plasma actuators can more effectively reduce the cavity tone by introducing streamwise vortices with the fine displacement.

The predicted phase-averaged flow fields show that the streamwise vortices by the control are introduced into the cavity flow as longitudinal vortices. Due to these vortices, large-scale vortices with spanwise axes that cause the cavity tone are distorted and weakened. Moreover, the coherence of the velocity fluctuations in the spanwise direction decreases, which also contributes to the reduction of the cavity tone.

Although the exact estimation of the power consumption is a future problem, the effectiveness of the plasma actuators as the devices for noise reduction can be shown in this paper. The technological problems related with high-voltage safety or high-frequency effects are also remaining challenges for application to actual vehicles.

Acknowledgments: This work was supported by JSPS KAKENHI grant number JP17K06153 and through the application development for Post K computer (FLAGSHIP 2020) by the Ministry of Education, Culture, Sports, Science, and Technology of Japan (MEXT). The authors appreciate Mr. Kusumoto for contributions throughout the manufacturing of the actuators.

Author Contributions: Hiroshi Yokoyama conceived and designed the experiments, conducted the simulations, and wrote the paper; Isamu Tanimoto performed the experiments; Hiroshi Yokoyama and Isamu Tanimoto analyzed the data; Isamu Tanimoto contributed reagents/materials/analysis tools; and Hiroshi Yokoyama and Akiyoshi Iida supervised the research.

Conflicts of Interest: The authors declare no conflict of interest. The founding sponsors had no role in the design of the study; in the collection, analyses, or interpretation of data; in the writing of the manuscript, and in the decision to publish the results.

References

1. Rossiter, J.E. *Wind-Tunnel Experiments on the Flow over Rectangular Cavities at Subsonic and Transonic Speeds*; No. 3438; Ministry of Aviation: New Delhi, India, 1964. Available online: http://naca.central.cranfield.ac.uk/reports/arc/rm/3438.pdf (accessed on 7 July 2017).
2. Powell, A. On the edge tone. *J. Acoust. Soc. Am.* **1961**, *33*, 395–409. [CrossRef]
3. East, L.F. Aerodynamically induced resonance in rectangular cavities. *J. Sound Vib.* **1966**, *3*, 277–287. [CrossRef]
4. Zhuang, N.; Alvi, F.S.; Alkislar, M.B.; Shih, C. Supersonic cavity flows and their control. *AIAA J.* **2006**, *44*, 2118–2128. [CrossRef]
5. Yokoyama, H.; Adachi, R.; Minato, T.; Iida, A. Experimental and Numerical Investigations on Control Methods of Cavity Tone by Blowing Jet in an Upstream Boundary Layer. *SAE Int. J. Passeng. Cars Mech. Syst.* **2017**, *10*, 13–21. [CrossRef]
6. Huang, X.; Zhang, X. Streamwise and spanwise plasma actuators for flow-induced cavity noise control. *Phys. Fluids* **2008**, *20*, 037101. [CrossRef]

7. Gupta, A.D.; Roy, S. Noise control of subsonic cavity flows using plasma actuated receptive channels. *J. Phys. D* **2014**, *47*, 1–5. [CrossRef]
8. Yokoyama, H.; Odawara, H.; Iida, A. Effects of freestream turbulence on cavity tone and sound source. *Int. J. Aerosp. Eng.* **2016**, *2016*, 7347106. [CrossRef]
9. Kaneda, I.; Sekimoto, S.; Nonomura, T.; Asada, K.; Oyama, A.; Fujii, K. An effective three-dimensional layout of actuation body force for separation control. *Int. J. Aerosp. Eng.* **2012**, *2012*, 786960. [CrossRef]
10. Lele, S.K. Compact finite difference schemes with spectral-like resolution. *J. Comput. Phys.* **1992**, *103*, 16–42. [CrossRef]
11. Rizzetta, D.P.; Visbal, M.R. Large-eddy simulation of supersonic cavity flow fields including flow control. *AIAA J.* **2003**, *41*, 1452–1462. [CrossRef]
12. Bogey, C.; Bailly, C. Large eddy simulations of round free jets using explicit filtering with/without dynamic Smagorinsky model. *Int. J. Heat Fluid Flow* **2006**, *27*, 603–610. [CrossRef]
13. Bogey, C.; Bailly, C. Turbulence and energy budget in a self-preserving round jet: Direct evaluation using large eddy simulation. *J. Fluid Mech.* **2009**, *627*, 129–160. [CrossRef]
14. Matsuura, K.; Kato, C. Large-eddy simulation of compressible transitional flows in a low-pressure turbine cascade. *AIAA J.* **2007**, *45*, 442–457. [CrossRef]
15. Gaitonde, D.V.; Visbal, M.R. Pade-type higher-order boundary filters for the Navier-Stokes equations. *AIAA J.* **2000**, *38*, 2103–2112. [CrossRef]
16. Yokoyama, H.; Kato, C. Fluid-acoustic interactions in self-sustained oscillations in turbulent cavity flows, I. Fluid-dynamic oscillations. *Phys. Fluids* **2009**, *21*, 105103. [CrossRef]
17. Suzen, Y.B.; Huang, P.G.; Jacob, J.D.; Ashpis, D.E. Numerical Simulations of Plasma Based Flow Control Applications. In Proceedings of the 35th Fluid Dynamics Conference and Exhibition, Toronto, ON, Canada, 6–9 June 2005; No. AIAA 2005-4633. pp. 1–11. [CrossRef]
18. Kusumoto, M.; Yokoyama, H.; Angland, D.; Iida, A. Control of aerodynamic noise from cascade of flat plates by plasma actuators. *Trans. JSME* **2017**, *83*, 16-00364. (In Japanese) [CrossRef]
19. Forte, M.; Jolibois, J.; Pons, J.; Moreau, E.; Touchard, G.; Cazalens, M. Optimization of a dielectric barrier discharge actuator by stationary and non-stationary measurements of the induced flow velocity: Application to airflow control. *Exp. Fluids* **2007**, *43*, 917–928. [CrossRef]
20. Thompson, K.W. Time dependent boundary conditions for hyperbolic systems. *J. Comput. Phys.* **1987**, *68*, 1–24. [CrossRef]
21. Poinsot, T.J.; Lele, S.K. Boundary conditions for direct simulations of compressible viscous flows. *J. Comput. Phys.* **1992**, *101*, 104–129. [CrossRef]
22. Kim, J.W.; Lee, D.J. Generalized characteristic boundary conditions for computational aeroacoustics. *AIAA J.* **1987**, *38*, 2040–2049. [CrossRef]
23. Williams, J.E.F.; Hawkings, D.L. Sound generation by turbulence and surfaces in arbitrary motion. *Philos. Trans. R. Soc. A* **1968**, *264*, 321–342. [CrossRef]
24. Lyrintzis, A.S. Surface integral methods in computational aeroacoustics—From the (CFD) near- field to the (Acoustic) far- field. *Int. J. Aeroacoust.* **2003**, *2*, 95–128. [CrossRef]
25. Shur, M.L.; Spalart, P.R.; Strelets, M.K. Noise prediction for increasingly complex jets. Part I: Methods and tests. *Int. J. Aeroacoust.* **2005**, *4*, 213–246. [CrossRef]
26. Yokoyama, H.; Kitamiya, K.; Iida, A. Flows around cascade of flat plates with acoustic resonance. *Phys. Fluids* **2013**, *25*, 106104. [CrossRef]
27. Enloe, C.L.; McLaughlin, T.E.; VanDyken, R.D.; Kachner, K.D.; Jumper, E.J.; Corke, T.C.; Post, M.; Haddad, O. Mechanisms and Responses of a Single Dielectric Barrier Plasma Actuator: Geometric Effects. *AIAA J.* **2004**, *42*, 595–604. [CrossRef]

© 2017 by the authors. Licensee MDPI, Basel, Switzerland. This article is an open access article distributed under the terms and conditions of the Creative Commons Attribution (CC BY) license (http://creativecommons.org/licenses/by/4.0/).

applied sciences

MDPI

Article

Combined CFD-Stochastic Analysis of an Active Fluidic Injection System for Jet Noise Reduction

Mattia Barbarino [1,*], Mario Ilsami [2], Raffaele Tuccillo [2] and Luigi Federico [1]

1 Department of Air Transport Environmental Impact, Italian Aerospace Research Center (CIRA), 81043 Capua, Italy; l.federico@cira.it
2 Department of Industrial Engineering, University of Naples "Federico II", 80138 Naples, Italy; mario.ilsami@gmail.com (M.I.); raffaele.tuccillo@unina.it (R.T.)
* Correspondence: m.barbarino@cira.it; Tel.: +39-0823-623-290

Academic Editor: Dimitrios G. Aggelis
Received: 3 May 2017; Accepted: 14 June 2017; Published: 16 June 2017

Abstract: In the framework of DANTE project (Development of Aero-Vibroacoustics Numerical and Technical Expertise), funded under the Italian Aerospace Research Program (PRORA), the prediction and reduction of noise from subsonic jets through the reconstruction of turbulent fields from Reynolds Averaged Navier Stokes (RANS) calculations are addressed. This approach, known as Stochastic Noise Generation and Radiation (SNGR), reconstructs the turbulent velocity fluctuations by RANS fields and calculates the source terms of Vortex Sound acoustic analogy. In the first part of this work, numerical and experimental jet-noise test cases have been reproduced by means RANS simulations and with different turbulence models in order to validate the approach for its subsequent use as a design tool. The noise spectra, predicted with SNGR, are in good agreement with both the experimental data and the results of Large-Eddy Simulations (LES). In the last part of this work, an active fluid injection technique, based on extractions from turbine and injections of high-pressure gas into the main stream of exhausts, has been proposed and finally assessed with the aim of reducing the jet-noise through the mixing and breaking of the turbulent eddies. Some tests have been carried out in order to set the best design parameters in terms of mass flow rate and injection velocity and to design the system functionalities. The SNGR method is, therefore, suitable to be used for the early design phase of jet-noise reduction technologies and a right combination of the fluid injection design parameters allows for a reduction of the jet-noise to 3.5 dB, as compared to the baseline case without injections.

Keywords: jet-noise; stochastic noise generation and radiation; Reynolds Averaged Navier Stokes; Large Eddy Simulations

1. Introduction

The problem of noise generation by compressible turbulent jets has been the subject of studies since the early 1950s, with the introduction of the turbojet also in commercial aircraft. It has continued, even later since the 1980s, with the introduction of turbofan with high bypass ratios, inherently less noisy than previously, due to the reduced exhaust velocity. In recent decades, the problem of jet noise prediction has been numerically addressed through a broad range of methods.

Despite the application of Direct Numerical Simulation (DNS) to jet-noise prediction [1–4], becoming more feasible with the growing advancement in computational resources, due to the large disparities of length and energy scales between fluid and acoustic fields, the use of fully solved Navier Stokes equations without turbulence modeling (DNS) is still restricted to low Reynolds number flows.

Instead, the numerical simulation of aeroacoustics through the solution of filtered Navier Stokes equations, either using fully Large Eddy Simulation (LES) or hybrid Reynolds Averaged Navier

Stokes-LES approaches, such as the detached eddy simulation (DES), is a major area of research [5–7]. However, despite the increase in computational power, even these types of simulations are not yet feasible for industrial purposes. Indeed, industry interest is mainly devoted to reliable numerical tools to be applied to realistic configurations for redesign of old configurations and for the development of new technologies. Furthermore, the growing interest on multi-disciplinary and multi-objective optimization necessarily lead to approaches that require low computational time.

Therefore, Reynolds Averaged Navier Stokes (RANS) simulations still remains the more feasible approach for Computational Fluid Dynamics (CFD) applications of industrial interest. However, RANS computations are not able to model, solely, the aeroacoustic phenomena. In this context, the stochastic approach for the prediction of noise from turbulence has received a great deal of interest in recent years. It was introduced by Kraichnan [8] and Fung [9], and it is based on the idea that Fourier components of solenoidal velocity fluctuations can be sampled in the wave-number space from a prescribed mono-dimensional energy spectrum. The revision and improvement of these methods for aeroacoustic applications have produced the stochastic noise generation and radiation (SNGR) methods [10–15].

Concerning the jet-noise reduction devices, in recent decades, chevron nozzles attracted much attention to their noise reduction benefits and are currently one of the most popular jet-noise reduction devices. Chevrons typically reduce low frequency noise, while increasing high frequency noise [16] with a limited thrust loss of only about 0.25%. Downstream vortex, generated by chevron, improves mixing in the shear layer, which leads to a decrease or increase in noise at certain frequency ranges. Several RANS computational studies were conducted about chevron nozzles [17–19]. The first LES calculations for the chevron nozzle jets appeared to be performed by Shur et al. [20,21].

A promising technology seems to be the fluidic chevron, as alternative solution to the mechanical chevron. This device consists of small injectors that inject high pressure air or other fluids, such as water, near the nozzle exit edge in order to emulate the mixing and the noise reduction features of the mechanical chevron. Numerous studies and experiments have been carried out to develop fluid chevron technology, especially at NASA Langley Jet Noise lab [22–24]. Fluidic chevrons, tested at NASA, were the first of their kind, designed with small slots near the nozzle exit edge to allow air injection into the stream and promote mixing between the core flow and the flow from the fan. Despite the simplicity of the mechanical chevron, the fluidic injection technique allows a greater flexibility. It can be used when the jet-noise reduction requirement is needed, for example during take-off and landing phases.

The main goal of this paper is two-fold. The first one is to make the assessment of an improved SNGR method based on the previous works of Casalino and Barbarino [13] and Di Francescantonio [14] through the comparison with experimental and LES data of a cold subsonic jet [7]. The second one is to assess the design of an active fluid injection technique, based on extractions from turbine and injections of high-pressure gas into the main stream of exhausts. The novelty of this paper is the proof of the potential of the SNGR approach as applied to the design of new devices suitable for jet noise reduction.

2. Model Description and Validation

2.1. Model Description

The SNGR approach assumes that the turbulent velocity field can be reconstructed as a summation of Fourier components, according to Kraichnan [8]. Therefore, the reconstructed turbulent velocity $\boldsymbol{u}'$ at point $\boldsymbol{x}$ and time t reads:

$$\boldsymbol{u}'(\boldsymbol{x},t) = 2\sum_{n=1}^{N_F} \hat{u}_n \cos\{\boldsymbol{k}_n \cdot (\boldsymbol{x} - \varrho \boldsymbol{U}t) + \psi_n\}\boldsymbol{\sigma}_n \quad (1)$$

where $\hat{u}_n$, ψ_n and $\boldsymbol{\sigma}_n$ are the magnitude, phase and direction of the nth Fourier component, respectively.

As proposed by Bailly and Juvé [11], each Fourier mode is supposed to be convected at the local mean-flow velocity $\boldsymbol{U}$ corrected by the vortex convection velocity ratio ϱ. This factor may account for the wall induction effect that reduces the vortex convection velocity with respect to the mean-flow velocity at the location of the vortex core. The factor may also account for the vertical induction in a jet shear layer, but this effect is negligible for low-speed subsonic jets. For jet noise prediction, the value $\varrho = 1$ has been used. Notice that the scalar product $\varrho \boldsymbol{k}_n \cdot \boldsymbol{U}$ accounts for the local time variation of the velocity field. Assuming incompressibility, the zero-divergence condition, applied to the Equation (1), results in the relationship $\boldsymbol{k}_n \cdot \boldsymbol{\sigma}_n = 0$, stating that the wave vector is perpendicular to the velocity vector.

By supposing that the turbulent flow field is isotropic, the magnitude of the nth Fourier mode is related to the mono-dimensional energy spectrum $E(k)$ by the expression, $\hat{u}_n = \sqrt{E(k_n)\Delta k_n}$, where k_n and Δk_n are the wave number and the corresponding band of the n-th mode. The Von Kármán—Pao isotropic turbulence spectrum is assumed [10]; that is:

$$E(k) = A(2/3)(K/k_e)(k/k_e)^4 exp\left[-2(k/k_\eta)\right]\left[1+(k/k_e)^2\right]^{-17/6} \tag{2}$$

where K is the turbulent kinetic energy, A is a numerical constant, k_e is the wave number of maximum energy, $k_\eta = \varepsilon^{1/4}\upsilon^{-3/4}$ is the Kolmogorov wave number, υ is the kinematical viscosity of the fluid and ε is the turbulent dissipation rate.

The constants A and k_e can be determined by equating the integral energy and the integral length scale derived from turbulence spectrum to the RANS quantities K and $L_T = c_1 u'^3/\varepsilon$, respectively, $u' = \sqrt{2K/3}$ being the isotropic turbulent velocity and c_1 the first tuning parameter of the method. The aforementioned calculation provides [11] $A = 1.453$ and $k_e = \frac{9\pi}{55}\frac{A}{L_T} = 0.747/L_T$.

The parameter c_1 allows tuning of the RANS turbulent integral length scale of the large-scale eddies. Its value is, by definition, close to unity, but its optimal value depends on the turbulent flow structure and conditions and on the RANS turbulence model. The optimal value of the parameter $c_1 = 2$ was found by Casalino and Barbarino [13], and used in this work.

The stochastic isotropic and homogeneous velocity perturbation field can be generated by choosing probability density functions for all the random variables involved in the Fourier decomposition [10]. These random variables are the angles φ_n, α_n, θ_n which define the direction of the wave vector $\boldsymbol{k}_n$, as sketched in Figure 1, the angle α_n, which defines the direction of the unit vector $\boldsymbol{\sigma}_n$ in a plane orthogonal to $\boldsymbol{k}_n$, and finally the phase angle ψ_n.

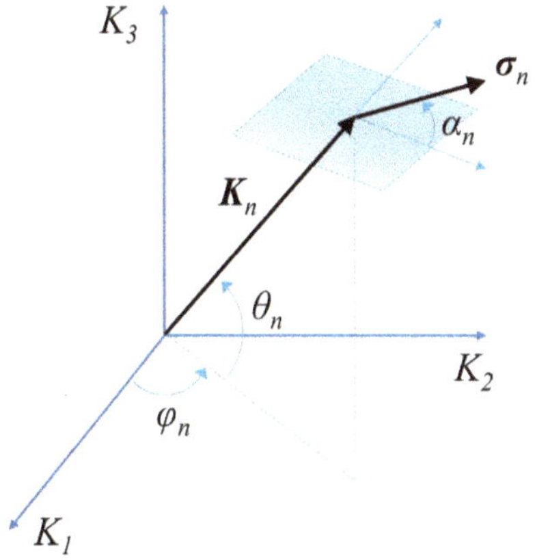

Figure 1. Representation of the wave vector $\boldsymbol{k}_n$ and velocity direction vector $\boldsymbol{\sigma}_n$ and definition of the stochastic angles (Adapted from [13]).

The requirement that the wave vector is uniformly distributed in the 3D wave-number space provides the following probability densities:

$$\mathrm{P}(\varphi_n) = (2\pi)^{-1}, \text{ with } -\pi \leq \varphi_n \leq \pi \tag{3}$$

$$\mathrm{P}(\theta_n) = cos(\theta_n)/2, \text{ with } -\pi/2 \leq \theta_n \leq \pi/2$$

Analogously, by supposing that the σ_n vector is uniformly distributed in the plane normal to k_n yields:

$$\mathrm{P}(\alpha_n) = (2\pi)^{-1}, \text{ with } -\pi \leq \alpha_n \leq \pi \tag{4}$$

Finally, the phase angle ψ_n is also supposed to be uniformly distributed in the 2π range, that is:

$$\mathrm{P}(\psi_n) = (2\pi)^{-1}, \text{ with } -\pi \leq \psi_n \leq \pi \tag{5}$$

The extremes of the wave number interval are related, at each node of the CFD mesh, to the value of the wave number k_e through the relationships:

$$k_1 = c_2 k_e;\ k_{N_F} = min\left(c_3 k_e, k_\eta, 2\pi/(N_\omega \Delta_i)\right) \tag{6}$$

where $\Delta_i = max\left(\left|U \cdot (x_j - x_i)\right|/U_i\right)$ is the maximum distance between node i and its neighboring stencil nodes j, projected in the direction of the local mean-flow velocity U_i; N_ω is the number of grid points for the Fourier component, set to 6 in the present work, as suggested by Bailly and Juvé [11]. The optimal value of the Fourier modes number, $N_F = 50$, was found by Casalino and Barbarino [13], and used in this work. The constants c_2 and c_3 are two additional tuning constants of the stochastic model: $c_2 = 0.1$, $c_3 = 10$.

The use of an adaptive wave-number interval allows reducing the number of Fourier modes. Furthermore, in order to obtain a better discretization of the energy spectrum at the lower energy-containing Fourier modes, a logarithmic distribution of the k_n values is used, say:

$$k_n = \exp[\ln(k_1) + D(n-1)];\ D = \left[\ln\left(k_{N_F}\right) - \ln(k_1)\right]/(N-1) \tag{7}$$

The energy spectrum $E(k)$ is integrated over each band n to obtain the magnitude of the velocity component $\hat{u}_n$ by discretizing the interval D in a number of constant subintervals, from the value $k_n^- = \exp[\ln(k_1) + D(n-1.5)]$ up to the value $k_n^- = \exp[\ln(k_1) + D(n-0.5)]$.

A crucial point of the SNGR approach is the definition of the two-point space correlation of the velocity fluctuations at each node of the CFD mesh by means of an artificial numerical approach able to reproduce the physical behaviour of the turbulent structures. The simplest approach, used for instance in Bechara [10], would consist of segmenting the bounding box of the active source region in square paths with edges equal to the average value of the correlation length over the whole source region. Then, a set of stochastic angles are sampled in each patch and these values are finally attributed to all the mesh nodes falling in the patch. The main drawback of this approach is that the correlation length is the same over all the source region, and therefore, does not account for the local Reynolds stresses and size of the turbulent structures.

Conversely, two different SNGR approaches, based on the the correlation length of the CFD solution, have been proposed. The first one consists in dividing the domain in blobs whose dimension is proportional to the correlation length of the CFD cells falling inside the blob. In the first step, the RANS correlation lengths $l_k^{\,i} = \left|-0.18(K^2/\epsilon)\partial U_k/\partial x_k + 2K/3\right|^{3/2}/\epsilon$ along the three Cartesian directions, each one denoted by the subscript k, are evaluated at each node, i, of the CFD mesh. Thus, random angles are assigned constant for each blob [13] as depicted in Figure 2a. The second one consists in projecting the CFD solution on an acoustic domain made of Cartesian cells whose dimension is proportional to the turbulent scale length $L_T = c_1 u'^3/\varepsilon$ of the CFD cells falling inside the acoustic

Cartesian cell (Figure 2b). The correlation length is assumed to be different for each Fourier mode of Equation (1) and proportional to the mode wavelength, $l^i = 3L_0/8$ where L_0 is the wavelength of each Fourier mode [14]. For each mode, random angles are assigned constant for the acoustic Cartesian cells having the same correlation length.

The CFD mesh, used in the first approach, is not isotropic because its cells are elongated. Therefore, the CFD solution must be projected on a hexahedral mesh with uniform cells, which greatly increases the number of sources and consequently also computational time and memory.

Conversely, the second approach uses an adaptive acoustic mesh, wherein the cell size is representative of the turbulent vortex size: within the stream there are small cells in correspondence of small vortices and there are large cells in correspondence of large vortices. This adaptive acoustic mesh is isotropic, i.e., its cells have the same dimensions in all three directions, but at the same time it is also variable, i.e., locally, cells size is proportional to the turbulent vortex length.

The main advantage of the second approach is the reduction of the total cells to be processed in respect of the CFD mesh, reducing computation time and memory but ensuring the required refinement.

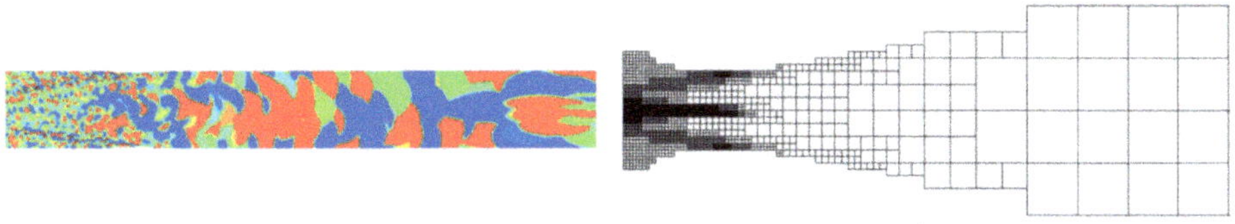

Figure 2. Two-point space correlation of a jet plume. (**a**) First approach based on blobs structure; (**b**) Second approach based on an adaptive Cartesian mesh.

The computed velocity field is finally transformed in the frequency domain and assemble to compute the frequency counter-part of the Lamb vector $(\boldsymbol{\omega}' \times \mathbf{u}')$.

According to the vortex sound analogy, the far field noise is finally achieved by the Powell's equation in frequency domain:

$$p = -\rho_0 \int_V (\boldsymbol{\omega}' \times \mathbf{u}') \cdot \nabla G dV \quad (8)$$

where G is the free-field Green function and $(\boldsymbol{\omega}' \times \mathbf{u}')$ is the frequency counter-part of the Lamb vector.

2.2. Model Validation

The SNGR approach has been firstly validated against experimental and LES results of a subsonic jet with diameter $D_j = 0.050$ m, Mach number $M = 0.75$ and the Reynolds number, $Re_D = 5.0 \times 10^4$, described by Andersson [7].

The two-dimensional geometry of the domain, shown in Figure 3, has been built using the CAD software Workbench by ANSYS.

A CFD RANS simulation based on the nozzle geometry used for the LES simulation [7] has been performed. The external profile of the nozzle has been reproduced by scanning from the original work [7]. The RANS simulation of the subsonic jet has been performed in the cold conditions, for which $T_j/T_\infty = 1$, where T_j is the static temperature of the jet at the nozzle outlet section, and T_∞ is the ambient temperature. In this study, the static temperature and pressure of the jet at the nozzle outlet section are respectively $T_j = 288$ K and $P_j = 101{,}300$ Pa. The temperature and pressure at the nozzle inlet section, respectively T_0 and P_0, were calculated using Rankine Hugoniot equations:

$$\begin{aligned} \frac{P_0}{P_j} &= \left(1 + \frac{k-1}{2} M^2\right)^{\frac{k}{k-1}} \\ \frac{T_0}{T_j} &= 1 + \frac{k-1}{2} M^2 \end{aligned} \tag{9}$$

where $M = 0.75$ is the Mach Number while $k = c_p/c_v = 1.4$ is the specific heat ratio of air, temperature and pressure at the nozzle inlet section, obtained by solving these two equations, are $P_0 = 147{,}100$ Pa and $T_0 = 320.4$ K.

The jet flow simulation has been carried out using the CFD software Fluent by ANSYS. The mesh extends from 0 to 50 nozzle diameters, D_j, in the axial direction. The radial extension is $10D_j$ at the nozzle outlet position and it increases up to $20D_j$ at the far-field outlet position. Mesh cells are refined in the region near the nozzle exit and in the jet shear layer.

A 2D axisymmetric transient pressure based second-order upwind scheme has been employed to converge fully coupled RANS equations with turbulence accounted for through K-ε and K-ω *SST* models. A view of the mesh made up of 1.85×10^5 cells and contour plots of the axial velocity and turbulent kinetic energy are shown in Figure 4.

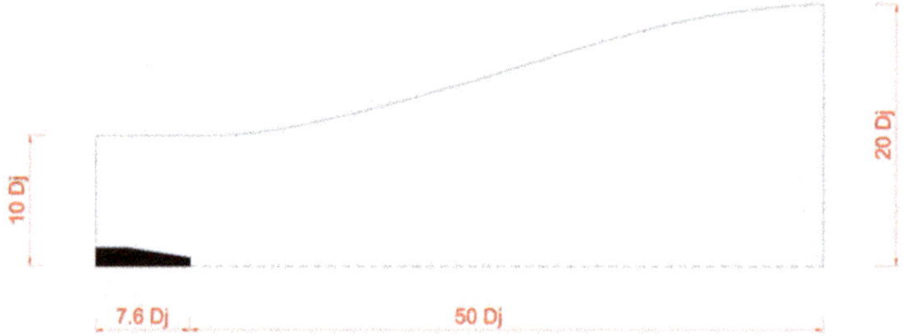

Figure 3. Computational domain (Adapted from Andersson [7]).

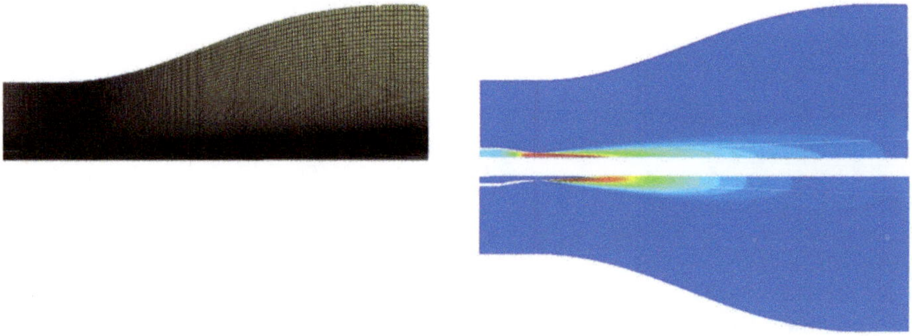

(a) **(b)**

Figure 4. (**a**) Mesh used for the Reynolds Averaged Navier Stokes (RANS) jet flow simulation; (**b**) Contour plots of the RANS solution. Mean velocity on the top and turbulent kinetic energy on the bottom.

The aerodynamic field results were extracted along the centerline and along radial lines at three axial positions downstream of the nozzle exit, according to Figure 5.

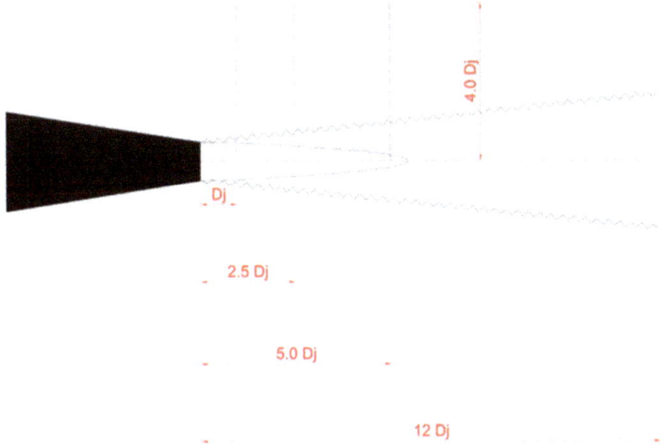

Figure 5. The vertical dashed lines indicate lines along which profiles of time-averaged quantities were extracted (Adapted from Andersson [7]).

The laminar core and the decay of the centerline velocity are in good agreement with the experimental results. The predicted laminar core length of $8D_j$ is slight higher than the experimental value as usual for RANS computations (Figure 6a). The maximum of centerline turbulent kinetic velocity occurs at about $10D_j$, in accordance with experimental results but its amplitude is slightly underestimated (Figure 6b).

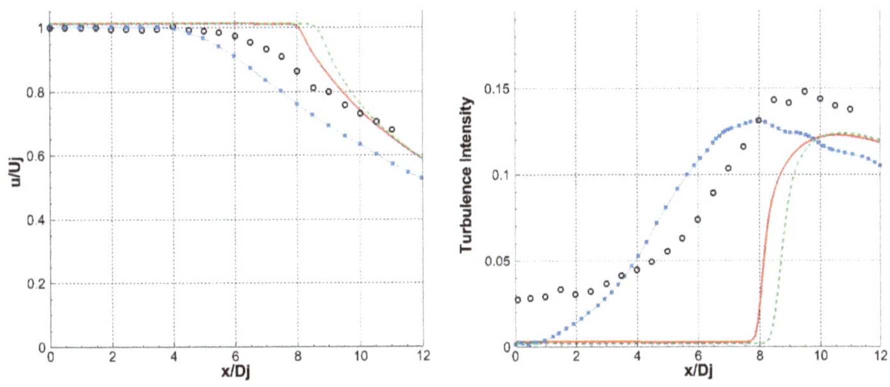

(a) (b)

Figure 6. (**a**) Centerline profile of the axial velocity; (**b**) Axial profiles of turbulence intensity. (Dotted black lines: experimental results; continuous red line: *K*-ε model; dashed green line: *K*-ω SST model; dotted-dashed blue line: LES).

Moreover, the radial profile of the axial velocity is well predicted by both RANS and LES computations (Figure 7a), whereas, the *uv* correlation levels of the radial velocity profile are underestimated by the RANS analysis due to lower mixing predicted and overestimated by the LES (Figure 7b).

It can be argued that the RANS analysis results are in fairly good agreement with experimental data and LES results although they underestimate the value of uv correlation and do not predict the gradual increase of turbulent energy in the region of the laminar core.

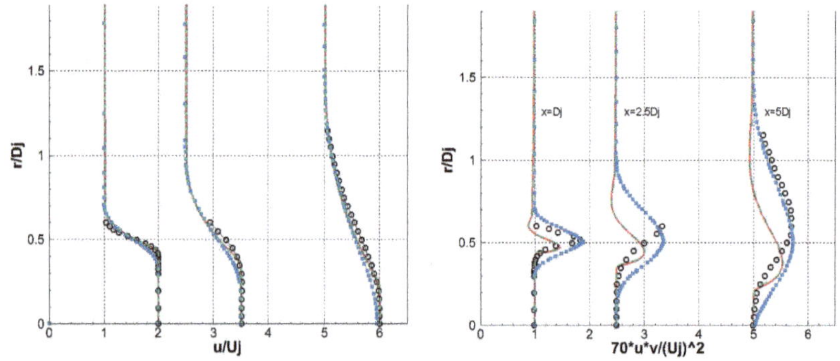

(a) (b)

Figure 7. (**a**) Radial profiles of axial velocity; (**b**) Radial profiles of uv correlation. The profiles have been staggered according to their axial location. (Dotted black lines: experimental results; continuous red line: K-ε model; dashed green line: K-ω SST model; dotted-dashed blue line: LES).

The acoustic field is evaluated by processing the RANS data by means the SNGR approach. To compute the stochastic noise sources, the axisymmetric CFD solution has been initially projected on a 3D domain, i.e., a uniform grid that consists of 1.74×10^6 cubic cells with $\Delta x = 5$ mm (Figure 8).

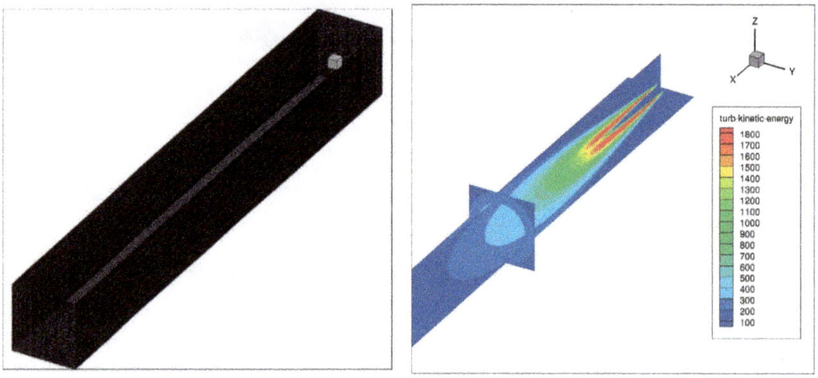

(a) (b)

Figure 8. (**a**) 3D acoustic domain; (**b**) 3D contour plot of turbulent kinetic energy.

The noise has been computed directly from the Fourier transform of the Lamb vector by using the integral solution of Powell's equation. Acoustic results are computed on a total of 25 microphones on two microphones arcs in the jet far-field region (Figure 9) and compared against LES and experimental results. Then, 14 microphones are arranged on a first arc of radius $30D_j$ from the center of the nozzle outlet section and positioned by 20 degrees to 150 degrees with an angular step of 10 degrees. The remaining 11 microphones were placed on a second arc of radius $50D_j$ from the center of the nozzle outlet section and positioned by 50 degrees to 150 degrees with an angular step of 10 degrees.

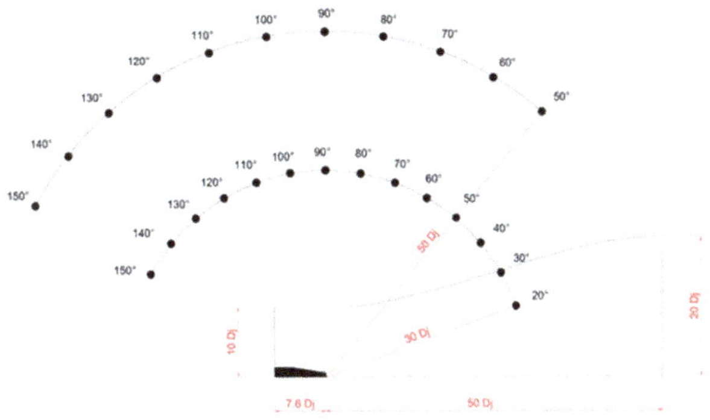

Figure 9. Position of the microphones (Adapted from Andersson [7]).

The acoustic results are expressed in terms of the PSD (Power Spectral Density) and OASPL (Over-All Sound Pressure Level).

To simplify the comparison with experimental data, the PSD spectra were filtered in third-octave bands. To represent all on the same graph, the PSD spectra have been staggered by multiplying the amplitude by a factor 10^{2n}, where $n = \left(\frac{\theta-20}{40}\right)$ and θ being the angle from the jet axis.

SNGR method does not take account for either convective effects or refractions by the shear layer. Therefore, SNGR results have been corrected, as suggested by Lighthill [15], to account for convective effects by using a Doppler factor of $1/(1 - M_c cos\theta)^{\alpha}$, where α is a proper exponential coefficient and M_c is the convective Mach number. Furthermore, in a first approximation, mean flow refraction effects have been neglected.

Figures 10 and 11 depict acoustic results for both arcs in terms of PSD spectra and OASPL levels showing that SNGR results are in very good agreement with the experimental results and LES analyses, with a matching within 2 dB except at the smaller angles where the maximum discrepancy is of about 5 dB (Figure 10b). In particular, the SNGR method allows predicting the trend of PSD spectra over a large part of the frequency range and up to $St = 1.5 \div 2$. Instead, the LES method allows a good prediction only up to $St = 1$; after that, the PSD spectra decay abruptly because of sub-grid filter that does not resolve the smallest scales, i.e., at the high frequencies. There are not great differences between the two models of turbulence, K-ε and K-ω SST.

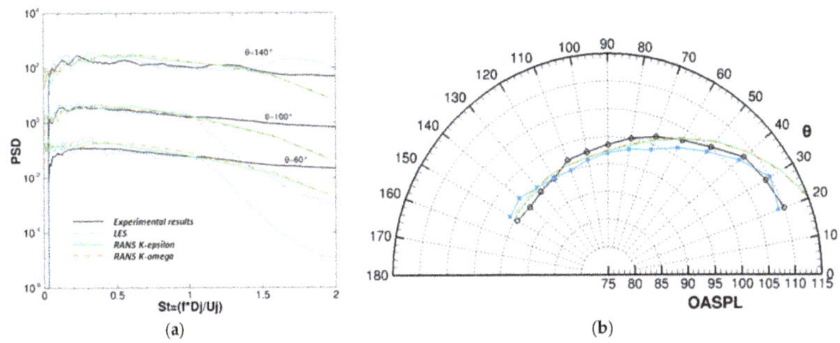

Figure 10. (**a**) Power spectra of far-field pressure signal for a few observer locations on the inner arc, $30D_j$. The Power Spectral Density (PSD) spectra have been staggered by multiplying the amplitude by a factor 10^{2n}, where $n = \left(\frac{\theta-20}{40}\right)$ and θ being the angle from the jet axis; (**b**) Overall Sound Pressure Level Directivity.

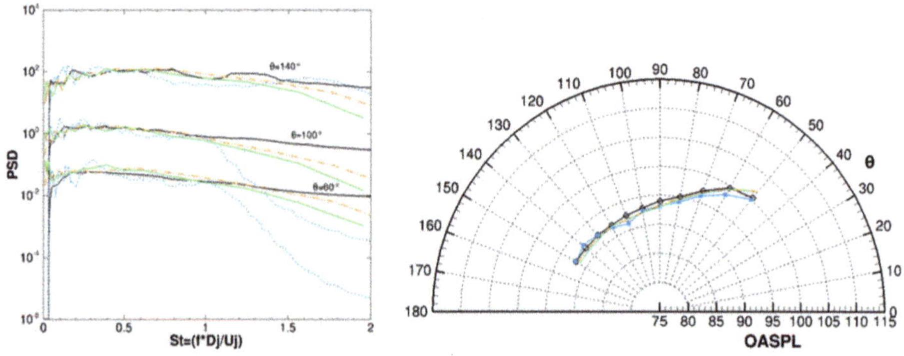

(a) (b)

Figure 11. (**a**) Power spectra of far-field pressure signal for a few observer locations on the outer arc, $30D_j$. The PSD spectra have been staggered by multiplying the amplitude by a factor 10^{2n}, where $n = \left(\frac{\theta-20}{40}\right)$ and θ being the angle from the jet axis; (**b**) Overall Sound Pressure Level Directivity.

In conclusion, the SNGR model, appropriately corrected by the Doppler factor, allows predicting the acoustic field of a subsonic jet both in terms of PSD spectra and in terms of directivity. In addition, the method is not very sensitive to the turbulence model adopted and allows estimating the jet-noise in a wide range of frequencies from a 2D axisymmetric RANS solution with a margin of uncertainty comparable to that of a LES calculation.

Finally, the contour plot of the real part of the acoustic pressure at 700 Hz and computed with the SNGR approach is shown in Figure 12. It can be observed that the lamb vector reconstructs the acoustic source encapsulated inside the jet flow region whereas the acoustic pressure propagates outside the jet region.

Figure 12. Sound radiation from the jet. Real part of the acoustic pressure [Pa] at 700 Hz.

3. Active Fluidic Injection System

In this section, an experimental technology for reducing the jet noise based on a fluidic injection is assessed by means of the SNGR approach and applied to the jet analyzed in the Section 2.2. Several experimental studies, conducted over recent decades, have shown that the fluid injection technique effectively allows reduction of the noise from turbulence [22–24]. Therefore, the technological solution proposed here consists of injecting a small jet of secondary gas into the main flow, which comes out from the nozzle of the jet engine exhausts (Figure 13). This secondary jet is derived from gas bleeding from the turbine located immediately upstream of the nozzle.

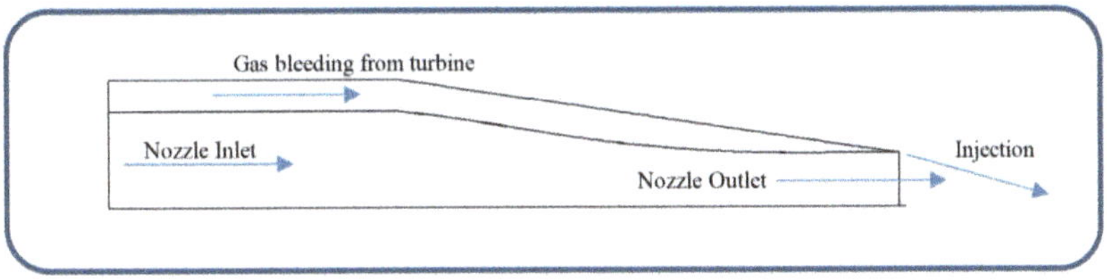

Figure 13. Sketch of the active Fluidic Injection System.

The air extractions from the turbine and subsequent injections in the main stream of the exhaust gas at nozzle outlet favor a greater mixing in order to break the large turbulent structures at low frequencies, which are the main responsible for most of the radiated noise (Figure 14).

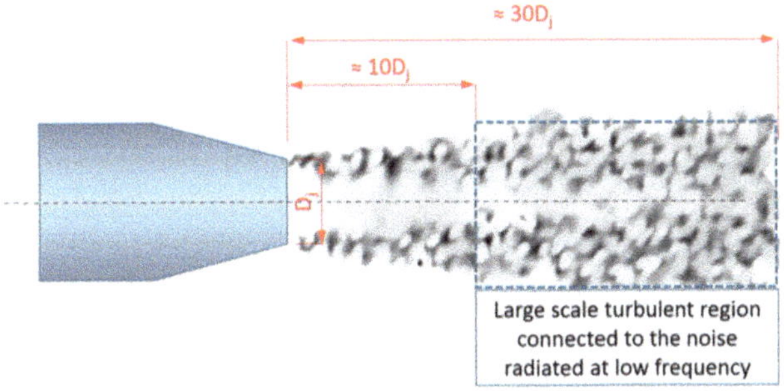

Figure 14. Sketch of the low frequency turbulence region.

Figure 15 shows the difference in the view of the exhaust plume emitted from a jet engine between the cases with and without fluid injection.

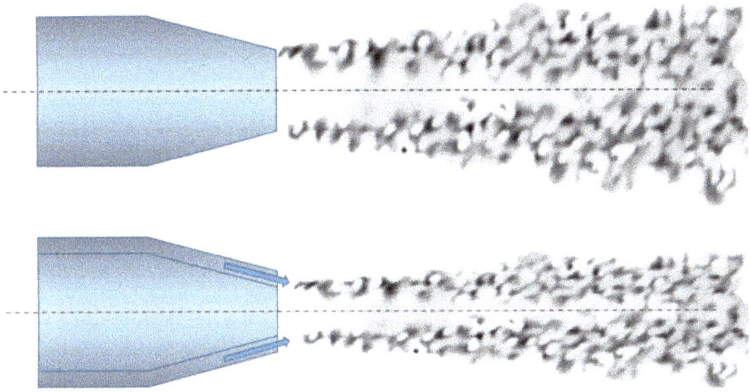

Figure 15. Sketch of the exhaust plumes emitted from a jet engine without (on the top) and with fluid injection (on the bottom).

In this work, four experimental tests were carried out to evaluate the influence of the bled mass flow rate, the area of the injection section and the jet angle over the noise reduction.

Two injection mass flow rate values are considered and corresponding about 5% and 10% of the total mass flow rate of the turbine. A maximum rate of about 10% has been used in order to preserve the nominal turbine power, whereas, pressure and temperature of the drained mass flow rate have been computed, assuming to be greater than the ambient conditions and according to the thermodynamic expansion. Two injection sections with a total area of around 4% and 10% of the nozzle outlet area have been considered and modelled in the computational domain and corresponding respectively to the 5% and 10% of the turbine mass flow rate. Finally, the spray conditions have been evaluated at three different injections angles equal to 45, 60 and 80 degrees.

The following test matrix (Table 1) has been, thus, assembled for the RANS + SNGR analyses. Examples of the different turbulence levels obtained by varying the injection patterns are displayed in Figure 16.

Table 1. Test matrix for the injection system design.

TEST	Injection Mass Flow Rate/Nozzle Mass Flow Rate	Injection Section Area/Nozzle Outlet Area	Injection Direction [deg]	Injection Velocity [m/s]
1	13.0%	10.2%	45	350
2	13.0%	10.2%	60	450
3	13.0%	10.2%	80	690
4	6.5%	4.08%	80	720

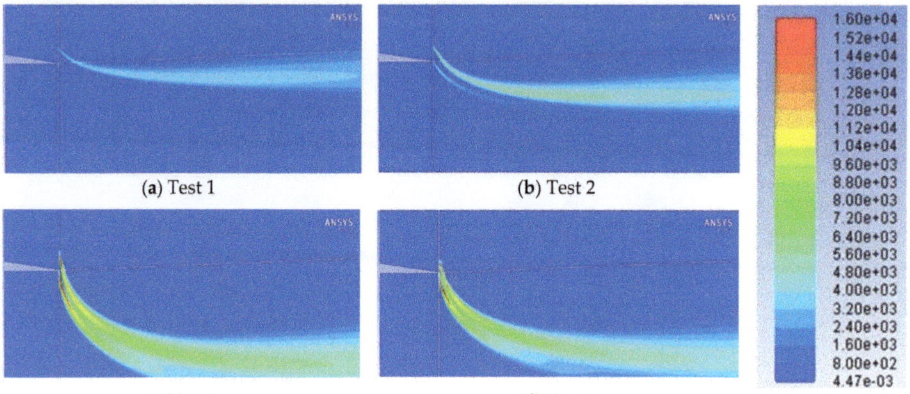

(**a**) Test 1 (**b**) Test 2 (**c**) Test 3 (**d**) Test 4

Figure 16. Turbulent kinetic energy levels [m^2/s^2] for the injection patterns tested.

About the aerodynamic field results, Figure 17a shows the centerline profiles of the axial velocity obtained from four tests with injection. Increasing the jet angle, the predicted laminar core length decreases compared to the baseline case. In particular, focusing on the results of 3rd and 4th tests, the laminar core length of 4th test is greater than the 3rd one, because at the same jet angle, $\gamma = 80°$, the injected jet has a higher speed, due to the lower injection section area. At the same time, increasing the jet angle, the intensity of turbulence profiles moves toward the nozzle exit, because the mixing is locally increasing (Figure 17b).

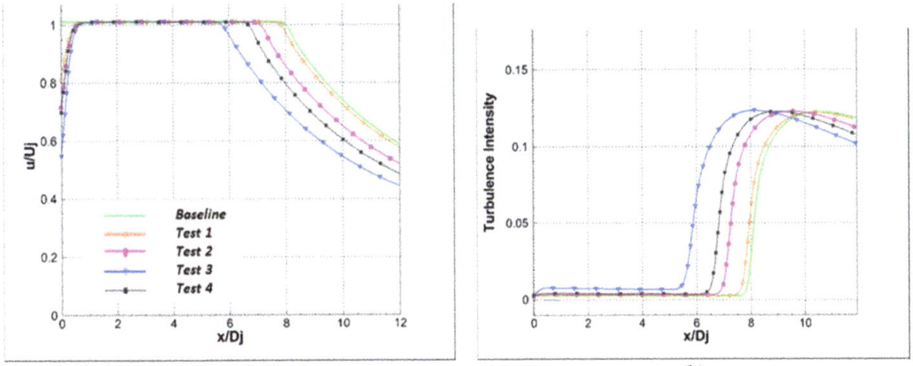

(a) (b)

Figure 17. Comparison between the different injection patterns in terms of velocity and turbulence intensity. (**a**) Centerline profile of the axial velocity; (**b**) Axial profiles of turbulence intensity.

Figure 18 shows turbulent velocity difference, Δu_{turb}, between the baseline case and the 4th test in the whole domain of interest. In the case with fluid injection, the impact between the injected jet and the main flow cause a locally rupture of the vortical structures. The obtained result is an increase of turbulence at nozzle outlet, i.e., at high frequencies, and a reduction of the turbulence in the downstream domain, which manifests itself as noise reduction at low frequencies.

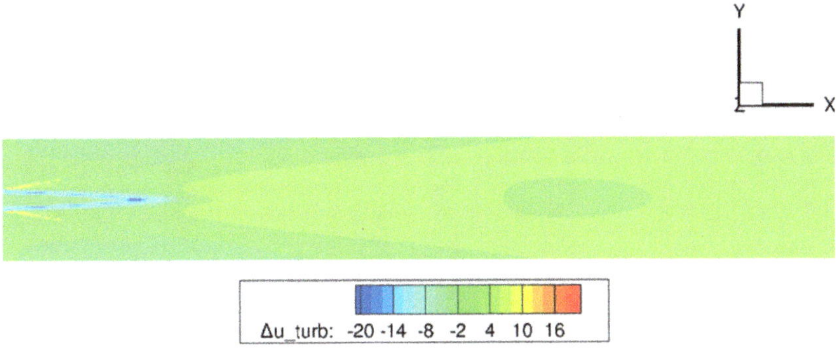

Figure 18. Turbulent velocity difference, Δu_{turb} [m/s] between the baseline case and the 4th test in the whole domain of interest.

Finally, Figure 19 shows the directivity patterns achieved with the tested fluidic injection system. It has been observed that the fluid injection technique produces a reduction of the OASPL compared to the baseline condition in all the four tests analyzed. In particular, in the fourth test condition the maximum reduction of the OASPL, of about 3.5 dB, is achieved at each directivity angle.

The experimental tests, carried out in this section, show that the fluid injection technique allows to reduce the jet-noise.

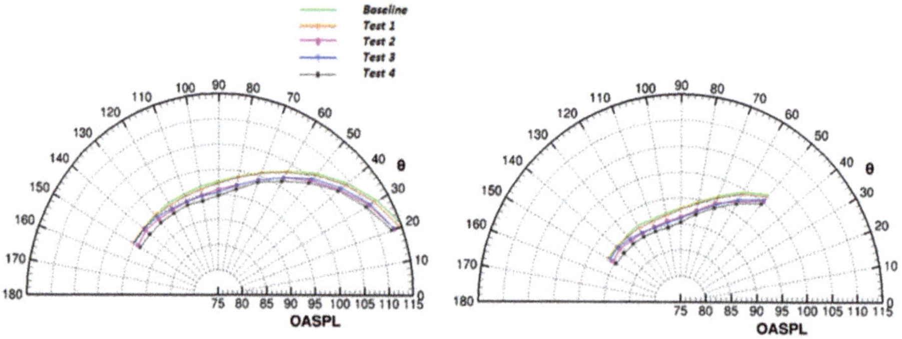

(a) (b)

Figure 19. (**a**) Overall Sound Pressure Level Directivity on the inner arc; (**b**) Overall Sound Pressure Level Directivity on the outer arc.

4. Conclusions

This paper illustrates the use of a computational aeroacoustics approach, named SNGR (Stochastic Noise Generation and Radiation), based on CFD RANS solutions, and aimed at jet-noise prediction and reduction.

The SNGR method, used in the present work, reconstructs the turbulent velocity fluctuations and calculates the source terms of Vortex Sound acoustic analogy equation, by using Green's free-field function.

In the first part of this work, the Andersson's test case has been reproduced to validate the coupled RANS–SNGR approach for its subsequent use as a design tool. The numerical results of aerodynamic and acoustic fields, obtained with the coupled RANS–SNGR approach, have been successfully assessed and validated against both experimental data and results of Large–Eddy Simulations (LES).

In the second part, a technique based on injecting small jets of secondary air, drained from the turbine, into the main jet flow is proposed and assessed. The fluid injection technique, unlike fixed and passive reduction systems such as mechanical chevrons, can be used in all cases that present a real need for noise reduction, such as takeoff and landing phases when the most intense phenomenon of jet noise takes place.

Four tests were carried out, involving gas bleed from the turbine and subsequent injections into the main flow of exhaust gas with the aim of aiding a greater mixing and breaking the large turbulent structures. The four tests have analyzed the influence on the noise reduction of several parameters, such as drained mass flow rate, the area of the injection section and the slope angle of the secondary jets at the same drained pressure and temperature. The best result was obtained in the 4th test, characterized by $\dot{m}_{spill}$ = 5% $\dot{m}_{tot}$, P_{spill} = 250,000 Pa, T_{spill} = 367.2 K, A_{inj} = 80 mm^2, γ = 80° and u_{inj} = 720 m/s, where the maximum noise reduction was achieved, by approximately 3.5 dB compared to Andersson's baseline case without injection.

Therefore, such tests have confirmed that the fluid injection technique allows an effective jet noise reduction. Moreover, like pointed out by the parametric study discussed in the previous section, the fluid injection technique can be designed as an active and adaptive system, after a proper optimization of several parameters, like the drained mass flow rate, the related pressure and temperature, the injection section area and jet angle, depending on the flight and engine operative conditions.

This kind of fluid injection technique will be controlled by an electronic control unit, in order to produce air outflows from the turbine, and then injections within the gas flow, only when needed. The electronic control of the fluid injection also should allow to regulate at any time the bled mass flow and the temperature and outlet pressure values, in order to optimize the noise reduction in accordance with the operating conditions. Finally, the efficacy of this technique can be further enhanced by azimuth control of the injections, which varies with the operating conditions.

Acknowledgments: This work was conducted as part of the framework of DANTE project (Development of Aero-vibroacoustics Numerical and Technical Expertise), funded under the Italian Aerospace Research Program (PRORA).

Author Contributions: Mattia Barbarino programmed and set the SNGR code for the numerical prediction of jet-noise and designed the noise reduction concept. Mario Ilsami performed the CFD simulations of the aerodynamic field and carried out the comparison with experimental data and LES results. Raffaele Tuccillo determined the features of the jet-noise reduction system and organized the numerical test campaign. Luigi Federico provided hints on SNGR applications and on techniques of noise reduction.

Conflicts of Interest: The authors declare no conflict of interest.

Nomenclature

A	numerical constant
A_{inj}	injection section area
c_1, c_2, c_3, c_4	stochastic model parameters
c_p, c_v	constant pressure and costant volume specific heat
D	nozzle diameter
E	monodimensional turbulent kinetic energy spectrum
G	Green function
K	turbulent kinetic energy
k	acoustic wave number
$\mathbf{k}$	turbulent wave vector
k_e	wave number of maximum E
k_η	Kolmogorov wave number
L_T	turbulence integral length scale
M	Mach number
M_c	convective Mach number
m	mass flow rate
N_F	number of Fourier modes
N_ω	number of grid points per Fourier component
p	acoustic pressure
$\mathcal{P}$	probability density function
P_0, T_0	jet pressure and temperature at the nozzle inlet section
P_j, T_j	jet pressure and temperature at the nozzle outlet section
P_∞, T_∞	ambient pressure and temperature
P_{spill}, T_{spill}	pressure and temperature of the drained mass flow rate
R	microphone radial distance
Re_D	jet Reynolds number based on nozzle diameter
St	Strouhal number
t	time
U	mean-flow velocity
U_j, M_j	jet centerline velocity and Mach number at nozzle outlet section
u', ω'	fluctuating turbulent velocity and vorticity
u_{inj}	injection velocity
$\hat{u}_n, \psi_n, \sigma_n$	magnitude, phase and direction of nth Fourier component of u'
α, φ, θ	turbulent wave vector random angles
γ	injection angle
Δ_i	maximum distance between node i and its neighboring nodes j

Δx	cubic cells dimension of 3D domain
Δu_{turb}	turbulent velocity difference
ε	turbulent dissipation rate
υ	kinematical viscosity
ϱ	vortex convection velocity ratio
ω	radian frequency or specific turbulent dissipation rate
t	time
U	mean-flow velocity
U_j, M_j	jet centerline velocity and Mach number at nozzle outlet section

References

1. Colonius, T.; Lele, S.K.; Moin, P. Sound generation in a mixing layer. *J. Fluid Mech.* **1997**, *330*, 375–409.
2. Colonius, T.; Mohseni, K.; Freund, J.B.; Lele, S.K.; Moin, P. Evaluation of noise radiation mechanisms in a turbulent jet. In Proceedings of the Summer Program of Center for Turbulence Research, Stanford, CA, USA, 2 June 1998; pp. 159–167.
3. Freund, J. Noise Sources in a Low-Reynolds-number Turbulent Jet at Mach 0.9. *J. Fluid Mech.* **2011**, *438*, 277–305. [CrossRef]
4. Mitchell, B.; Lele, S.; Moin, P. Direct Computation of the Sound Generated by Vortex Pairing in an Axisymmetric Jet. *J. Fluid Mech.* **1999**, *383*, 113–142. [CrossRef]
5. Bogey, C.; Bailly, C.; Juvè, D. Computation of the Sound Radiated by a 3-D Jet Using Large Eddy Simulation. In Proceedings of the 6th AIAA/CEAS Aeroacoustics Conference, Lahaina, HI, USA, 12–14 June 2000.
6. Mankbadi, R.; Shih, S.; Hixon, R.; Povinelli, L. Direct Computation of Jet Noise Produced by Large-Scale Axisymmetric Structures. *J. Propuls. Power* **2000**, *16*, 207–215. [CrossRef]
7. Andersson, N.; Eriksson, L.E.; Davidson, L. A Study of Mach 0.75 Jets and Their Radiated Sound Using Large-Eddy Simulation. In Proceedings of the 10th AIAA/CEAS Aeroacoustics Conference, Manchester, UK, 10–12 May 2004.
8. Kraichnan, R.H. Diffusion by a Random Velocity Field. *Phys. Fluids* **1970**, *13*, 22–31. [CrossRef]
9. Fung, J.C.H.; Hunt, J.C.R.; Malik, N.A.; Perkins, R.J. Kinematic Simulation of Homogeneous Turbulence by Unsteady Random Fourier Modes. *J. Fluid Mech.* **1992**, *236*, 281–318. [CrossRef]
10. Béchara, W.; Bailly, C.; Lafon, P.; Candel, S. Stochastic Approach to Noise Modeling for Free Turbulent Flows. *AIAA J.* **1994**, *32*, 455–464. [CrossRef]
11. Bailly, C.; Juvé, D. A Stochastic Approach to Compute Subsonic Noise Using Linearized Euler's Equations. In Proceedings of the 5th AIAA/CEAS Aeroacoustics Conference, Seattle, WA, USA, 8–9 November 1999.
12. Billson, M.; Eriksson, L.; Davidson, L. Jet Noise Modeling Using Synthetic Anisotropic Turbulence. In Proceedings of the 10th AIAA/CEAS Aeroacoustics Conference, Manchester, UK, 10–12 May 2004.
13. Casalino, D.; Barbarino, M. Stochastic Method for Airfoil Self-Noise Computation in Frequency-Domain. *AIAA J.* **2011**, *49*, 2453–2469. [CrossRef]
14. Di Francescantonio, P. Side Mirror Noise with Adaptive Spectral Reconstruction. In Proceedings of the SAE 2015 Noise and Vibration Conference and Exhibition, Grand Rapids, MI, USA, 22–25 June 2015.
15. Lighthill, M.J. Jet Noise. *AIAA J.* **1963**, *1*, 1507–1517. [CrossRef]
16. Bridges, J.; Brown, C.A. Parametric Testing of Chevrons on Single Flow Hot Jets: NASA TM 2004–213107. In Proceedings of the 10th AIAA/CEAS Aeroacoustics Conference, Manchester, UK, 10–12 May 2004.
17. Engblom, W.A.; Kharavan, A.; Bridges, J. Numerical Prediction of Chevron Nozzle Noise Reduction Using WIND-MGBK Methodology. In Proceedings of the 10th AIAA/CEAS Aeroacoustics Conference, Manchester, UK, 10–12 May 2004.
18. Birch, S.F.; Lyubimov, D.A.; Maslov, V.P.; Secundov, A.N. Noise Prediction for Chevron Nozzle Flows. In Proceedings of the 12nd AIAA/CEAS Aeroacoustics Conference, Cambridge, MA, USA, 8–10 May 2006.
19. Massey, S.J.; Elmiligui, A.A.; Hunter, C.A.; Thomas, R.H.; Pao, S.P.; Mengle, V.G. Computational Analysis of a Chevron Nozzle Uniquely Tailored for Propulsion Airframe Aeroacoustics. In Proceedings of the 12nd AIAA/CEAS Aeroacoustics Conference, Cambridge, MA, USA, 8–10 May 2006.
20. Shur, M.L.; Spalart, P.R.; Strelets, M.K. Noise Prediction for Increasingly Complex Jets, Part II: Applications. *Int. J. Aeroacoustics* **2005**, *4*, 247–266. [CrossRef]

21. Shur, M.L.; Spalart, P.R.; Strelets, M.K.; Garbaruk, A.V. Further Steps in LES-Based Noise Prediction for Complex Jets. In Proceedings of the 44th AIAA Aerospace Sciences Meeting and Exhibit, Reno, NV, USA, 9–12 January 2006.
22. Henderson, B.S.; Kinzie, K.W.; Whitmire, J.; Abeysinghe, A. The impact of fluidic chevrons on jet noise. In Proceedings of the 11th AIAA/CEAS Aeroacoustics Conference, Monterey, CA, USA, 23–25 May 2005.
23. Kinzie, K.W.; Henderson, B.S.; Whitmire, J.; Abeysinghe, A. Fluidic chevrons for jet noise reduction. In Proceedings of the 2004 International Symposium on Active Control of Sound and Vibration (ACTIVE 2004), Williamsburg, VA, USA, 20–22 September 2004.
24. Henderson, B.S.; Kinzie, K.W.; Whitmire, J.; Abeysinghe, A. Aeroacoustic Improvements to Fluidic Chevron Nozzles. In Proceedings of the 12th AIAA/CEAS Aeroacoustics Conference, Cambridge, MA, USA, 8–10 May 2006.

© 2017 by the authors. Licensee MDPI, Basel, Switzerland. This article is an open access article distributed under the terms and conditions of the Creative Commons Attribution (CC BY) license (http://creativecommons.org/licenses/by/4.0/).

applied sciences

MDPI

Article

Surrogate Based Optimization of Aerodynamic Noise for Streamlined Shape of High Speed Trains

Zhenxu Sun *, Ye Zhang and Guowei Yang

Key Laboratory for Mechanics in Fluid Solid Coupling Systems, Institute of Mechanics, Chinese Academy of Sciences, Beijing 100190, China; zhangye@imech.ac.cn (Y.Z.); gwyang@imech.ac.cn (G.Y.)

* Correspondence: sunzhenxu@imech.ac.cn; Tel.: +86-10-8254-3815

Academic Editors: Roberto Citarella and Luigi Federico

Received: 3 January 2017; Accepted: 13 February 2017; Published: 17 February 2017

Abstract: Aerodynamic noise increases with the sixth power of the running speed. As the speed increases, aerodynamic noise becomes predominant and begins to be the main noise source at a certain high speed. As a result, aerodynamic noise has to be focused on when designing new high-speed trains. In order to perform the aerodynamic noise optimization, the equivalent continuous sound pressure level (SPL) has been used in the present paper, which could take all of the far field observation probes into consideration. The Non-Linear Acoustics Solver (NLAS) approach has been utilized for acoustic calculation. With the use of Kriging surrogate model, a multi-objective optimization of the streamlined shape of high-speed trains has been performed, which takes the noise level in the far field and the drag of the whole train as the objectives. To efficiently construct the Kriging model, the cross validation approach has been adopted. Optimization results reveal that both the equivalent continuous sound pressure level and the drag of the whole train are reduced in a certain extent.

Keywords: aerodynamic noise; NLAS; kriging model; cross validation; equivalent continuous SPL

1. Introduction

As the running speed of high-speed trains increases, aerodynamic noise becomes to be predominant and takes the major part of train noise. Severe noise problem will cause discomfort for the passengers and surrounding people. Furthermore, it will also cause damage to surrounding equipment and buildings. Noise issue becomes a key factor that limits the increase of running speed of high-speed trains, which needs to pay enough attention in the design of new trains [1].

Current study on aerodynamic noise of high-speed trains mainly focuses on two aspects: On the one hand, identification of noise sources of high-speed trains and mechanism analysis [2–6] were performed, aiming at obtaining more precise acoustic structures of high-speed trains. Talotte presents a critical survey of the identification and modeling of railway noise sources and summarizes the current knowledge of the physical source phenomena [2,3]. Sun numerically analyzes the main noise sources of high-speed trains, quantitatively compares the different noise sources and identifies the main noise sources, such as the leading streamline, the inter-coach spacing, the bogie zone and the wake zone as well [4,6]. Noger performed an investigation on acoustic production mechanisms in the TGV (which is short from the French words "train à grande vitesse") pantograph recess to be able to reduce the radiated noise [5]. Until now, the study on aerodynamic noise mechanism has gradually matured. On the other hand, many studies focus on noise reduction of specific components to meet practical usage [6–10]. In order to obtain low noise pantograph, Ikada proposes a precise rail pantograph design method that provides both aerodynamic and aero-acoustic sound designs more quickly and economically [7,8]. West Japan Railway Co. made an investigation to find an optimum shape of the front bulkhead of the pantograph cover and at the same time actively pursues development of a wing-shaped pantograph which causes less aerodynamic noise. Iwamoto and Higashi made a technical view from results of their tests using the actual railcar and the wind test tunnel [9]. Sueki proposed a new aerodynamic noise reduction method that involves covering the surface of objects with a particular porous material [10]. Sun investigated on different designs of cab windows and cowcatchers and gave a reasonable advice on low noise design of the streamlined shape [6]. It can be seen that much effort has been put to the design of low-noise pantograph, which engineering experience is mostly dependent on for noise reduction. This kind of design is usually constrained by engineering experience and cannot obtain the most optimal shape. In addition, study on the noise reduction of the streamlined head of high-speed trains could be seldom referred to, since abundant flow phenomena exist on the surface of the streamline and the design approach which relies solely on engineering experience could not meet the accuracy requirement. Consequently, optimization design with corresponding algorithms is the only way to obtain the relationships between design variables and aerodynamic noise and gets the chance to obtain the optimal design.

In recent years, optimization algorithms have been widely used in the streamlined shape design of high-speed trains [11–16]. Lee and Kim dealt with the nose shape design of high-speed railways to minimize the maximum micro-pressure wave and suggested an optimal nose shape that is an improvement over current design in terms of micro-pressure wave [11]. Vytla used the Kriging model to find the optimum shape of a two-dimensional nose shape of a high-speed train traveling at 350 km/h considering both the induced aerodynamic drag and the generated aerodynamic noise [12]. Ku used the VMF parameterization method on the train nose to reduce a micro-pressure wave and an aerodynamic drag of a high-speed train [13]. Yao performed a constrained multi-objective optimization design on a simplified CRH380A high-speed train with three carriages to optimize aerodynamic drag of the total train and aerodynamic lift of the trailing car and got agreeable optimal results [14]. Suzuki and Nakade presented a novel design technique for high-speed trains using a multi-objective optimization method to balance plural aerodynamic properties [15]. Muñoz-Paniagua used genetic algorithms to optimize the nose shape of a high-speed train entering a tunnel in term of the compression wave generated at the entry of the train and the aerodynamic drag of the train [16]. It could be concluded that the optimization objectives are mostly focused on the aerodynamic loads or tunnel effects. Even aerodynamic noise is considered, it is only a two-dimensional study. Due to the massive cost of computational aero-acoustics (CAA), aerodynamic noise optimization always seems untouchable for researchers. In order to perform aerodynamic noise optimization, two problems have to be solved in advance: On the one hand, the efficiency of CAA calculation has to be improved, which means proper CAA algorithm should be determined to adapt the computing scale for high-speed trains; On the other hand, assessment of the noise level in the far field should be determined. Once the above problems have been solved, the aerodynamic noise optimization becomes feasible. It is very crucial to construct an aerodynamic noise optimization strategy for practical streamlined shape of high-speed trains.

In the present paper, the authors present an aerodynamic noise optimization strategy for practical streamlined shape of high-speed trains. The NLAS approach [17] has been adopted to perform CAA analysis for each design sample. Meanwhile, the equivalent continuous SPL has been introduced to take all of the far field observation probes into consideration and taken as one objective for multi-objective optimization. With the use of Kriging surrogate model, the aerodynamic noise optimization of the streamlined shape of high-speed trains could be performed. In the present study, the noise level in the far field and the drag of the whole train are taken as the objectives. It is mainly based on the following considerations: The aerodynamic drag of high-speed trains can be up to 75% of the total drag at the speed of 300 km/h and the drag characteristics of the trains are directly related to the ability of energy saving and environmental protection. As a result, the aerodynamic drag reduction is very prominent and has become one of the key issues for aerodynamic shape optimization of high-speed trains. Meanwhile, aerodynamic noise increases with the sixth power of the running speed, and would be predominant over mechanical noise when the running speed is over 250 km/h. It is found that simply reducing aerodynamic drag cannot efficiently reduce the aerodynamic noise, indicating that aerodynamic noise optimization has to take aerodynamic noise as one optimization objective.

After optimization, aerodynamic performance of the optimized shape and the prototype shape with three carriages is comparatively analyzed. Finally, sensitivity analysis has been performed, so that the nonlinear relationship between the objectives and design variables could be obtained. Current research pushes forward the cognition on how to reduce the noise level by modifying the streamlined shape of high-speed trains.

2. Algorithms

2.1. Acoustic Algorithms and Validation

Two kinds of approaches are adopted in the present work, NLAS [17] to solve the near field noise and FW-H sound propagation method to solve the far field noise, respectively. For the latter approach, an acoustic surface around the noise sources is built to record fluctuation data during NLAS calculation, which would be taken as the initial value for FW-H propagation equation. The overall solving process is shown in Figure 1:

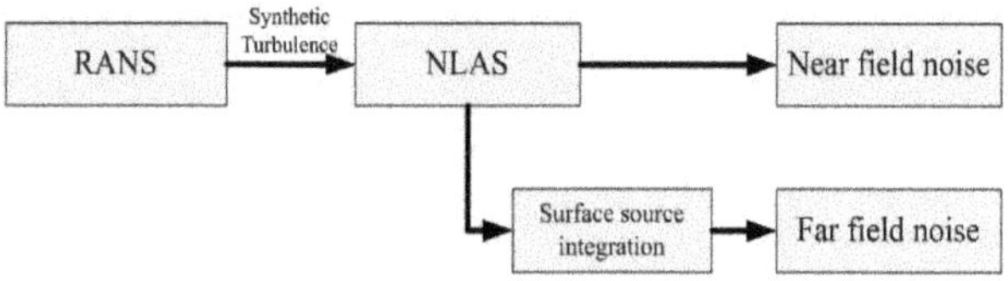

Figure 1. The schematic drawing of solution procedure for Computational aero-acoustics (CAA) calculation.

The NLAS is a numerical acoustics solver designed to model noise generation and propagation from an initial statistically-steady model of turbulent flow data, which can be provided by a simple Reynolds Averaged Navier-Stokes (RANS) turbulence model. Through the RANS calculation, the statistical steady mean flow could be obtained, from which the main generation zone of turbulence could be revealed. In the present paper, an anisotropic turbulence model, the cubic k-ε model, is utilized to preferably model the statistically steady flow field [18]. Non-linear terms are taken into consideration to account for normal-stress anisotropy, swirl and streamline curvature effects. Consequently, the local Reynolds-stress tensor could be provided precisely to synthesize the noise sources. The transient calculation of NLAS should be performed on the premise of the steady RANS calculation of the flow field. NLAS uses a reconstruction procedure to generate noise sources from the given set of statistics and allows the resulting propagation of the pressure disturbances to be simulated using a high-resolution pre-conditioned solver.

The acoustic surface is used to record fluctuation data during the NLAS procedure. Once the data is obtained, Ffowcs-Williams/Hawking equation can be solved based on these data. The FW-H equation approach [19,20] can be used to predict noise at any observation point outside the acoustic surface, even if the observation point is outside the computational domain. The whole CAA analysis is performed by the commercial software CAA++ (Ver 10.1, Metacomp Technologies, Los Angeles, CA, USA, 2011).

Numerical validation has already been performed in the literature by the author [4]. It took a backward step case as the test case. It comes from the experiment by Lee and Sung in a subsonic wind tunnel [21]. The flow conditions in the numerical simulation maintain the same with the experiment. In the experiment, the span wise width is 12.5 times of the height of the backward step to keep the central section two-dimensional. As a result, a two-dimensional simulation was conducted for the central section. The observation points are chosen just on the floor behind the step, as the experiment did. Fast Fourier Transformation (FFT) analysis is performed on the pressure fluctuating data. Meanwhile, numerical simulation with the Large Eddy Simulation (LES) approach was also performed. The results from the NLAS approach, the LES approach and the experiment are listed together for comparison, which mainly focuses on the sound pressure level and corresponding dominant frequency (Tables 1 and 2).

Table 1. Maximal sound pressure level (decibel, dB), where x is the distance from the probe to the step, and H is the height of the step.

Positions	$x/H = 2$	$x/H = 4$	$x/H = 6$	$x/H = 8$	$x/H = 10$
experiment	−26	−27	−24.2	−22.6	−24
NLAS	−26	−23	−24.35	−21.7	−23.2
LES	−26.27	−25.2	−23.04	−18.6	−19.54

NLAS: Nonlinear Acoustics Solver; LES: Large Eddy Simulation.

Table 2. Dominant frequency (Hz), where x is the distance from the probe to the step, and H is the height of the step.

Positions	$x/H = 2$	$x/H = 4$	$x/H = 6$	$x/H = 8$	$x/H = 10$
experiment	11.5	10.5	18	12	18
NLAS	14	7.1	13	13	9.1
LES	13.9	14.5	28.5	28.5	43

It can be seen that better dominant frequencies are achieved through the NLAS approach, except for the position $x/H = 10$. Meanwhile, the maximal sound pressure level predicted by NLAS agrees well with the experimental data. However, the maximal sound pressure level and dominant frequency obtained by LES under the same mesh and flow conditions show relative variation with experimental data. This can be related to that the synthetic reconstruction of sub-grid sources can be achieved in NLAS approach rather than LES approach, which makes the results from NLAS approach seems much more accurate. Simulation results reveal that the NLAS approach is an efficient and high-resolution computational method for aerodynamic noise prediction, and can be adopted for the large-scale computation of aerodynamic noise generated by high-speed trains.

2.2. *Assessment of the Noise Level in the Far Field*

When performing the CAA calculation, the train is assumed stationary, while the air passes by the train. Consequently, acoustic probes are placed equidistantly along the train to perform noise analysis. Due to the relatively large number of acoustic probes, it is difficult to assess the noise level by only one or two probes among them. When the acoustic optimization is conducted, it is necessary to take all the acoustic probes into consideration. Considering that the real circumstance is that the train passes by the probes one by one with the same speed, the equivalent continuous sound pressure level (SPL) is introduced in the present paper, which takes the form as:

$$L_{Aeq} = 10\lg\left[\frac{1}{T}\int_0^T 10^{0.1L_A}dt\right] \quad (1)$$

where T is the measurement time and L_A is the A-weighted SPL at time t.

As long as the noise probes are placed equidistantly, the equivalent continuous SPL could be discretized as shown below:

$$L_{Aeq} = 10\lg[\frac{1}{n}\sum_{i=1}^{n} 10^{\frac{L_{Ai}}{10}}] \quad (2)$$

where n is the number of the acoustic probes and L_{Ai} is the A-weighted SPL at the probe i.

Since the equivalent continuous SPL could takes all the acoustic probes in the far field into consideration, here it is taken as the objective to represent the noise level in the far field for the whole train.

2.3. Algorithms for Computational Fluid Dynamics (CFD) and Validation

As one goal of the whole optimization process, the drag of the whole train has to be evaluated, which is performed by CFD analysis. CFD accuracy directly affects the construction of the surrogate model and efficiency of optimization algorithm. In this paper, the speed of high-speed train is 300 km·h^{-1}. Under this condition, air compressibility has to be considered. Therefore, the steady compressible Reynolds-averaged Navier–Stokes equations [22] based on the finite volume method are used to predict the aerodynamic drag. The k-ω SST model [23] is selected as the turbulence model. The standard wall functions are used near the wall so that the accuracy of the CFD results could be ensured with a limited amount of mesh.

To validate the CFD algorithms, and to obtain a reasonable mesh configuration for aerodynamic simulation of high-speed trains, the authors performed a comparison study with wind tunnel experimental results. The experiments were performed in China Aerodynamics R&D Center. The wind tunnel model is a 1:8 scaled train model, with the bogies and inter-coach spacing included, as shown in Figure 2a. The velocity of the air flow is 60 m/s. The computational domain is the same as the tunnel walls. Due to the complex geometry, local densification of the mesh is performed, and the total number of grids is about 21.7 million. The mesh configuration is shown in Figure 2b.

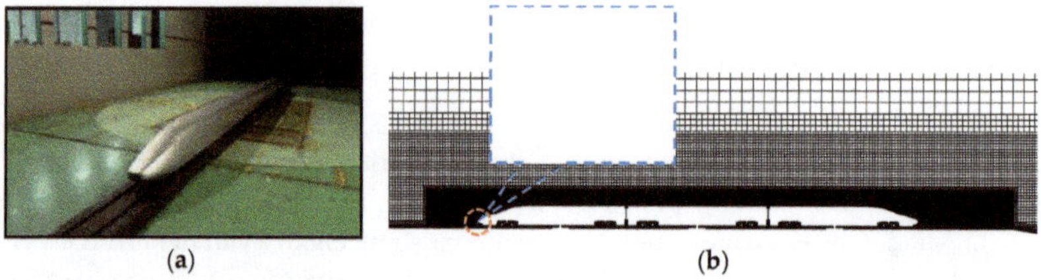

Figure 2. Wind tunnel train model and mesh configuration for numerical simulation: (**a**) wind tunnel model; and (**b**) mesh configuration.

Table 3 shows the comparison of aerodynamic coefficients between experimental results and numerical results. It can be seen that the aerodynamic coefficient error for each car are all within 5%, which is acceptable for engineering application, indicating that the CFD algorithms and mesh configuration for the simulation are both feasible for aerodynamic drag evaluation of high-speed trains, and will be utilized in the present study.

Table 3. Comparison of aerodynamic coefficients between experimental results and numerical results.

Aerodynamic Loads	Total-Cd	Head-Cd	Middle-Cd	Tail-Cd
Experiment	0.326	0.125	0.082	0.119
CFD	0.310	0.118	0.079	0.113
Error	4.91%	5.60%	3.66%	5.00%

CFD: Computational Fluid Dynamics.

3. Local Shape Function Parametric Approach

In this paper, a parametric approach called Local Shape Function (LSF) which is based on FFD method [24] and NURBS method [25] has been designed.

The whole processes are as follows:

(1) For a given geometry, deformation regions should be divided firstly.
(2) Mesh the deformation regions, and obtain the coordinate values of every grid point. In order to keep the smooth transition of the surface, the structural grids have been utilized for mesh discretization, as shown in Figure 3.

(3) Choose the deformation function of each region, which can be selected randomly, but smooth transition between adjacent regions should be ensured.
(4) Choose a weight factor W_i for each deformation function, which determines the maximum deformation value of each region.
(5) Calculate the increments Δ of coordinates of all grid points by the deformation functions and W_i.
(6) Get the coordinates of the deformed shape by summing Δ and the coordinates of the original shape.
(7) According to the coordinates of the deformed shape, the deformed surface can be fitted exactly, then a deformation process is done.

In the above process, Step (3) is the most crucial. The deformation surfaces are different from each other due to the different choices of deformation functions. Inappropriate deformation functions will easily lead to irrational deformation surfaces. Trigonometric functions, exponential functions, logarithmic functions, polynomial functions and NURBS functions are all commonly used for deformation functions.

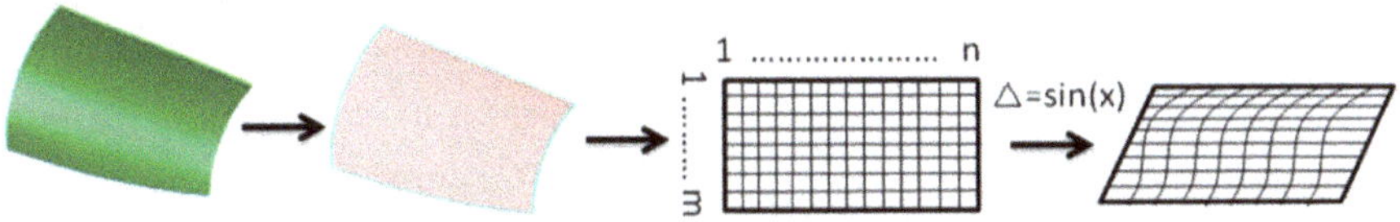

Figure 3. Schematic of surface deformation by Local Shape Function (LSF) method.

Due to the symmetrical design along the longitude of the train, only one side of the symmetrical plane of the streamline is parameterized. As a result, the design parameters can be reduced by half. The parametric surfaces are separated into four deformation regions, as shown in Figure 4a. The width of the streamline is controlled by Zone1, and a control point Point1 is set here, extracting its y coordinate as the third design parameter w_1. The slope of the cab window is controlled by Zone2, and the second control point Point2 is set here, extracting its z coordinate as the fourth design parameter w_2. Nose height is controlled by Zone3, and another control point Point3 is set here, extracting its z coordinate as the second design parameter w_3. Nose drainage is controlled by Zone4, and a control point Point4 is set here, extracting its y coordinate as the first design parameter w_4. For simplicity, all the deformation functions in this paper are trigonometric functions. Figure 4b shows the deformation in the height of the nose and the cab window. As seen above, the deformation method can ensure the surface smoothness and smooth transition among different deformation regions.

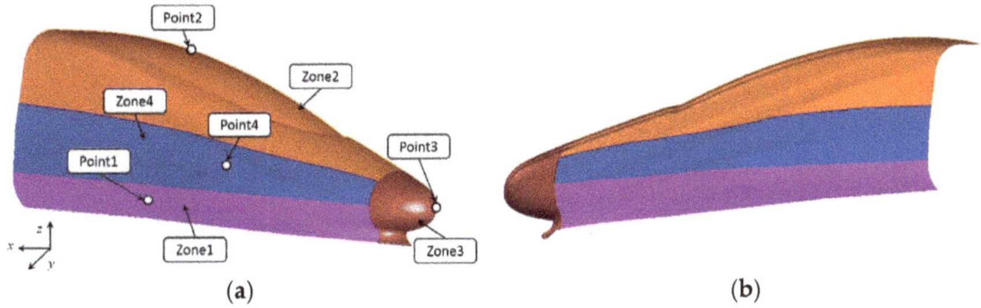

Figure 4. Deformation zones and local deformations: (**a**) schematic of deformation zones; and (**b**) the deformation of the nose and the cab window.

4. Optimization Strategy

4.1. Construction of Cross-Validation Based Kriging Model

Considering the huge computational cost in the aerodynamic optimization, especially in the shape optimization of high-speed trains, a sequential optimization method based on minimizing the response surface criteria has been adopted in the present paper to reduce the training points. The Kriging surrogate model is an interpolation technique based on statistical theory [26–29]. This model takes full account of the relevant characteristics of the variable space, containing the regression part and the nonparametric part. The Kriging model is obtained based on the cross-validation algorithm [30]. The use of cross-validation algorithm could prolong the time of the training process. However, compared to the massive cost of CFD and CAA calculation for each sampling point, this time cost could be ignorable. A basic Kriging model could be firstly constructed from the initial 20 training points. Based on this model, the Pareto set could be obtained. Two typical points of the Pareto set could be chosen for CFD/CAA validation. If the optimization accuracy could not meet the requirement, these two points then would be added to the initial training points to build a more accurate Kriging model. If the optimization accuracy meets the design requirement, the Kriging model is finally constructed. During each iteration process, the training points are divided into groups randomly, and each group has two training points.

4.2. Optimization Process

Figure 5 shows the whole optimization process designed in present paper. First, the Latin hypercube method is adopted to initialize the sampling points in the design space. Then, the CFD analysis is performed to obtain the values of the objectives corresponding to the sampling points. Third, the Kriging model is constructed via genetic algorithm (GA) technique based on these sampling points. After the Kriging model has been constructed, the multi-objective optimization based on self-adaptive GA approach could be performed and the Pareto set could be obtained. Last, the final Kriging model and multi-objective self-adaptive GA approach are both adopted to obtain the Pareto solutions in the design space.

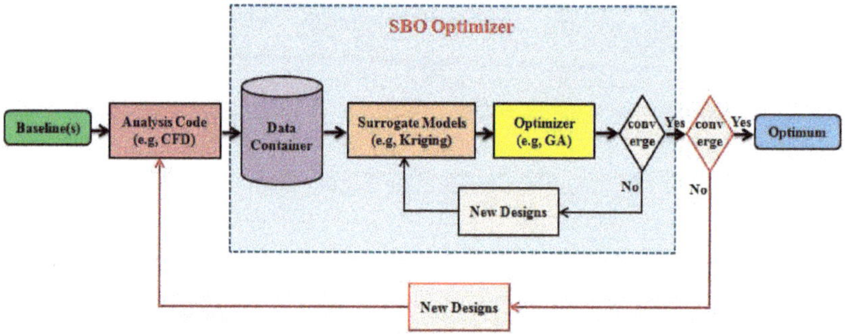

Figure 5. Schematic drawing of surrogated based optimization.

Based on the above process, multi-objective optimization of the streamlined shape of high-speed trains is performed, which will be discussed in the following sections.

5. Computational Models, Mesh and Conditions

A three-grouping high-speed train model has been adopted in the present paper, which is named as EMU1. The total length of the train is about 78 m and the height of the train is about 4.05 m. The whole train is shown in Figure 6.

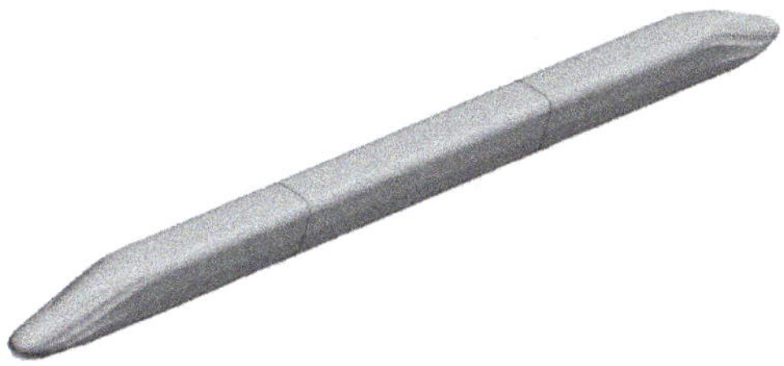

Figure 6. The whole train model.

In order to reduce the computational cost, some additional components are eliminated, such as the pantograph. When performing the CFD calculation, components such as the windshields and bogies are reserved. However, since the computational cost of CAA analysis is extremely larger than that of CFD, the windshields and bogies are both neglected to reduce the calculation time, yet the influence of the streamlined shape on acoustic characteristics could still be considered.

Take the height of the train as the characteristic length H, the computational domain extends 30 H ahead of the train nose and 60 H from the train tail to the exit of the computational domain. The top of the computational domain is at a distance 30 H from the bottom of the rail and the sides are at a distance of 30 H from the center axis of the train, the outline of computational domain and the model are shown in Figure 7.

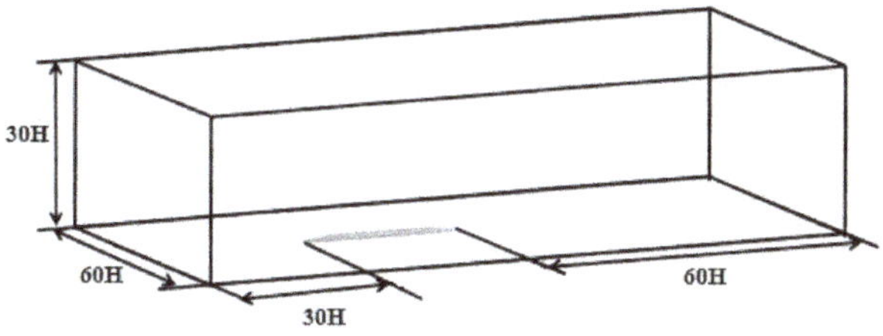

Figure 7. Computational domain.

The hybrid Cartesian/prism grids is adopted and six layers of prism grids are generated with an increasing ratio of 1.2 and a total length of 30 mm, which keeps the value of $y+$ of the first layer near the train surface in a range of 30~100. The total number of the cells is about 32 million. Figure 8 shows the grids on the longitudinal section and on the surface of leading streamline, while Figure 9 shows the grids on the longitudinal section of wake region.

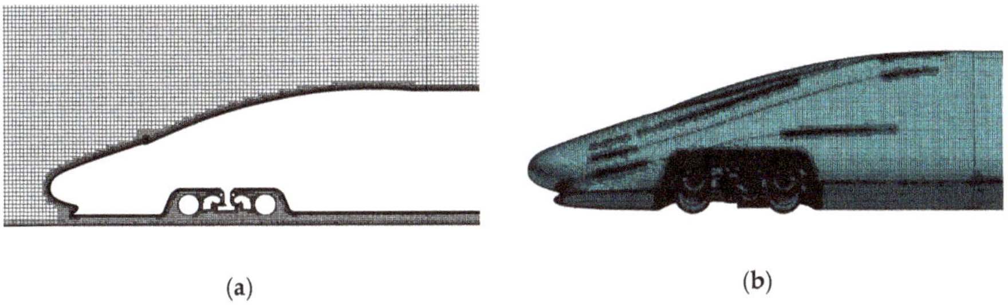

Figure 8. The grids on the longitudinal section and on the surface of leading streamline: (**a**) mesh distribution on the longitudinal section; and (**b**) surface grids of the leading streamline.

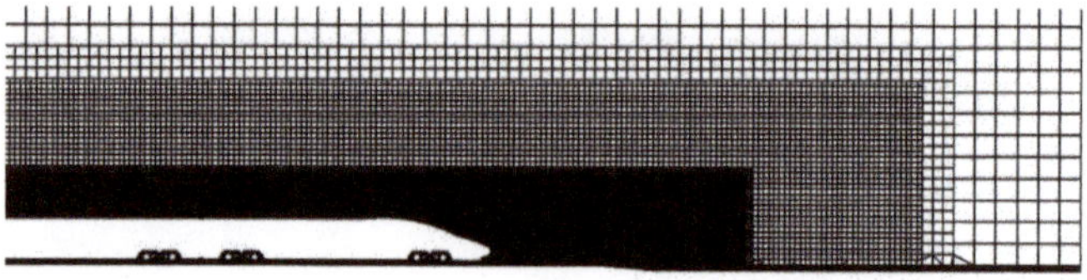

Figure 9. The grids on the longitudinal section of wake region.

The flow velocity is 300 km/h; the far-field pressure is 1 atm; the temperature is 288 K; and the reference area is the maximum cross-sectional area of the train. As a result of the compressibility calculation model, one-dimensional inviscid flow of the Riemann invariants is introduced as the far-field boundary conditions, which are also known as non-reflective boundary conditions. Inflow, outflow and the top boundaries are all set as far-field boundary conditions and the train body is non-slip solid wall boundary condition. The ground is treated as the moving wall to simulate the ground effect, and the moving speed is equal to the train speed.

When the NLAS procedure is conducted, three absorbing layers are imposed to the inlet boundary, the outlet boundary and the far field boundary respectively to prevent wave reflections from these boundaries. Two hundred Fourier modes are set to perform synthetic reconstruction for the turbulent fluctuating quantities to capture the sub-grid sources correctly. The time step in NLAS simulation is set to $2 \times e^{-5}$ s, and the simulated physical time is 0.3 s, which insures that the noise whose frequencies locate between 10–10,000 Hz could be precisely predicted.

After solving the acoustic field in the near field, the far field noise could be obtained with the use of FW-H equation. The observation probes are set according to the ISO-2005-3095 standard, which requires the probes to be 25 m far from the train body and 3.5 m height from the ground. Thirteen probes are placed along the train body with an equal distance of 10 m, which range from the leading head to the trailing streamline, as shown in Figure 10.

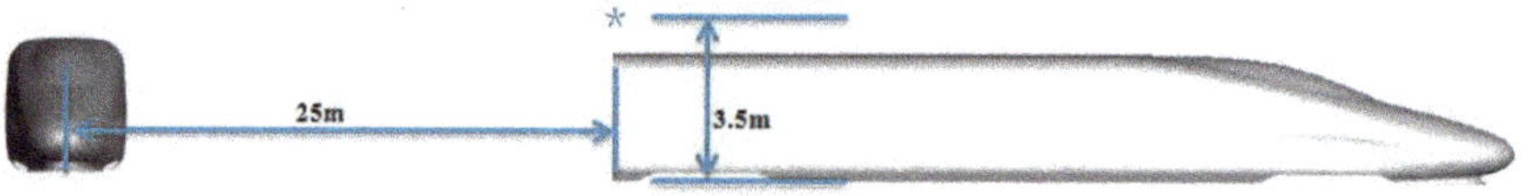

Figure 10. Schematic drawing of the observation probes in the far field.

6. Results and Discussion

As parameters for GA operation, the size of initial population is set to 100. The probability of crossover is 0.9 while the probability of mutation is set to 0.3. The size of evolution generations is set to 100. The value of θ varies from 0 to 10. After adding points for three times, the Kriging model meets the accuracy requirement. Figure 11 shows the convergence history of the fitness and the exact value for each variable.

The total aerodynamic drag coefficient and the equivalent continuous SPL are treated as the optimization objectives, and the multi-objective optimization of the streamlined shape has been performed on the final built Kriging model. The adaptive genetic approach with a population number of 200 is adopted. Three thousand generations have been performed in the optimization process. The roulette method is used as the selection operator, while the probabilities of crossover and mutation are set as 0.9 and 0.3, respectively. Figure 12 shows the Pareto set of the two objectives. It can be concluded that the optimal solutions of the objectives are both limited in a small zone, indicating that specific individual in the Pareto set shows no large differences, while all the individuals are better than the initial one. In order to demonstrate the mechanism for aerodynamic performance improvement, a specific individual is chosen randomly in Figure 12 as an example, just as the red spot shows.

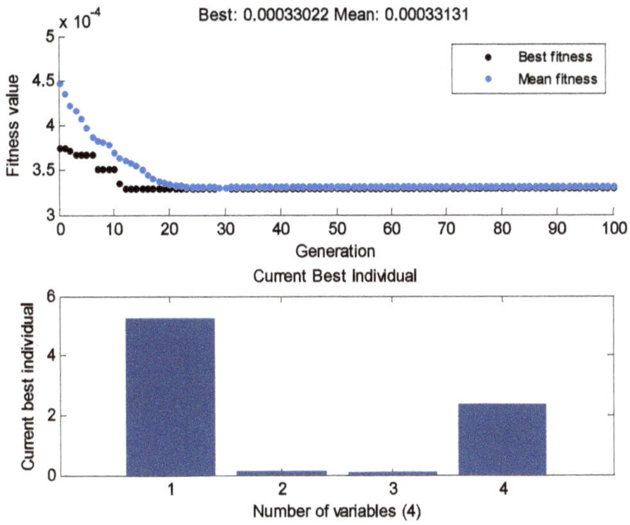

Figure 11. The convergence history of the fitness and the exact value for each variable.

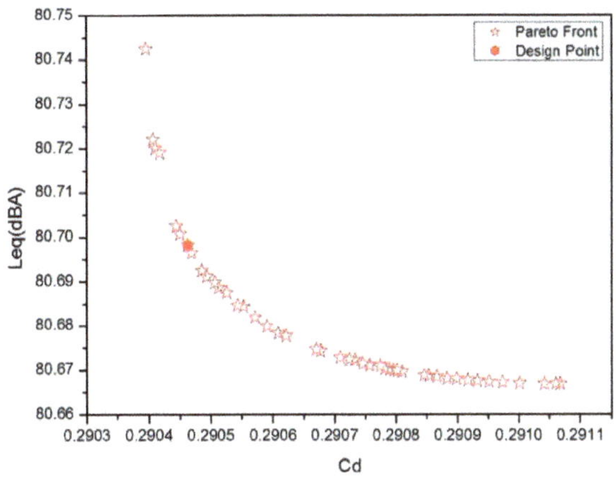

Figure 12. The Pareto front based on the drag of the whole train and the equivalent continuous SPL.

The drag coefficient of the whole train and the equivalent continuous SPL for the prototype of EMU1 are 0.314 and 81.93 dBA, respectively, while the objectives in the Pareto set are all more optimal than those of the prototype.

Table 4 shows the comparison between the CFD calculations and prediction from the Kriging model. It can be seen that excellent agreement exists between the two approaches. The errors for both objectives are 0.493% and 0.027%, respectively, which can meet the engineering requirement. It can be concluded that the Kriging model built in the present paper could reveal the relationships between the design variables and the objectives.

Table 4. The comparison between the CFD calculations and prediction from the Kriging model.

Objectives	Real Leq (dBA)	Kriging (dBA)	Error	Real Cd	Kriging
Test Case	80.676	80.698	0.027%	0.2919	0.29046

Figure 13 shows the variation of the streamlined shape between the prototype and the optimal result:

Figure 13. Comparison of the streamlined shape between the prototype and the optimal shape. (**a**) View from the direction of positive *y*; and (**b**) view from the direction of negative *y*.

As shown above, the green one represents the prototype and the orange one represents the optimal shape. Good smoothness can be observed on the shape after optimization. The width of the bottom of the streamline and the shape of the cowcatcher keep unchanged, while the width of the upper part of the streamline decreases and the heights of the nose and the cab window are both lowered.

Table 5 shows the variations of the design variables compared to the prototype model. W_1 controls the width of the streamline, W_2 controls the height of the cab window, W_3 controls the height of the nose and W_4 controls the width of the drainage. It can be seen that all the design variables get smaller.

Table 5. The variations of the design variables compared to the prototype model (1:1).

Variations	W_1/mm	W_2/mm	W_3/mm	W_4/mm
Test Case	−40.85	−38.50	−50.81	−61.55

Table 6 shows the comparison of the objectives before and after optimization. Results reveal that the aerodynamic performance gets improved for the optimal shape. The drag coefficient and the equivalent continuous SPL are reduced by 7.1% and 1.26 dB, respectively. For the optimization of aerodynamic drag, reducing the inviscid drag plays an important role. As seen in Table 3, the inviscid drag is reduced by 17.33%.

Table 6. The comparison of the objectives before and after optimization.

Objectives	Leq (dBA)	Cd	Cd_inviscid
Original	81.9316	0.314114	0.165
Test Case	80.676	0.2919	0.1364
Reduction	1.5%	7.1%	17.33%

In order to better understand the aerodynamic performance before and after optimization and investigate on the influence on the specific car by the deformation of streamlined head, the drag coefficients of each carriage before and after optimization are given in Figure 14:

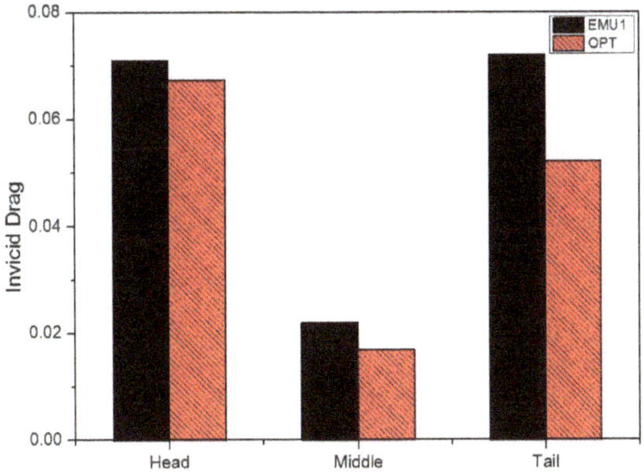

Figure 14. Comparison of drag coefficients of each carriage before and after optimization.

As shown above, the inviscid drag mainly exists on the leading car and trailing car. The inviscid drag is reduced by 5.2% and 27.6% for the leading car and trailing car, respectively. Meanwhile, the inviscid drag on the middle car is also reduced. The inviscid drag is the main source for drag optimization.

The pressure contour on the leading car before and after optimization is shown in Figure 15.

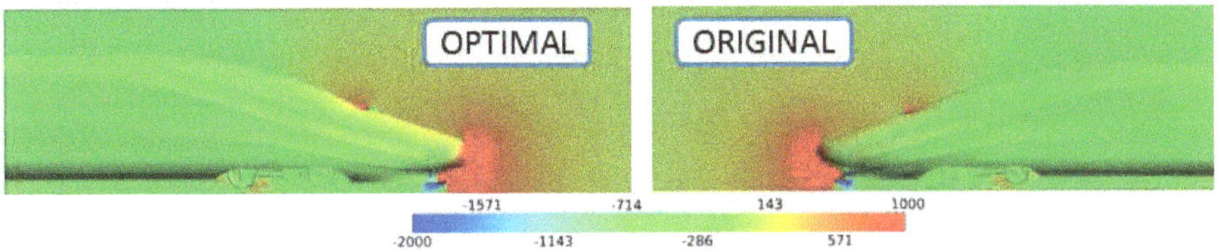

Figure 15. Pressure contour on the leading car before and after optimization (Pressure: Pa).

The height of the nose is reduced for the optimal model. Consequently, the high-pressure zone in front of the nose varies. The high-pressure zone mainly covers the region between the nose and the cowcatcher. Lower nose leads to a relatively smaller high-pressure zone, which results in a relatively smaller inviscid drag.

The pressure contour on the trailing car before and after optimization is shown in Figure 16.

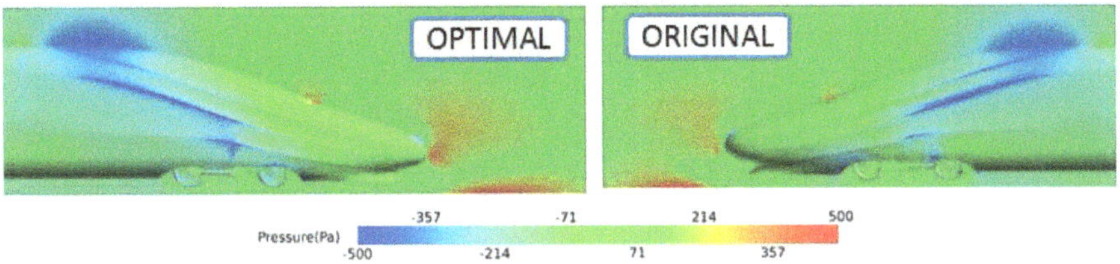

Figure 16. Pressure contour on the trailing car before and after optimization.

An obvious higher-pressure zone behind the trailing nose could be observed for the optimal model, which could provide a thrust on the trailing car. As a result, the inviscid drag on the trailing train could be effectively reduced.

The pressure coefficient along the longitudinal profile of the leading car and trailing car before and after optimization is also investigated, as shown in Figure 17.

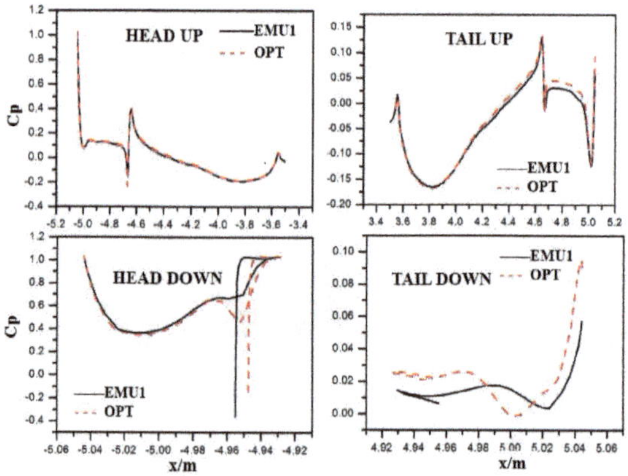

Figure 17. Pressure coefficient along the longitudinal profile of the leading car and trailing car before and after optimization.

It can be seen that the biggest variation exists at the bottom part of the trailing nose. Higher pressure could be observed along the lower longitudinal profile for the optimal model, which is the root for the relatively lower inviscid drag of the trailing car.

Then, the acoustic characteristics in the far field are analyzed. As mentioned above, the equivalent continuous SPL is reduced by 1.26 dB. Figure 18 shows the A-weighted overall sound pressure level (OASPLA) of all the probes before and after optimization.

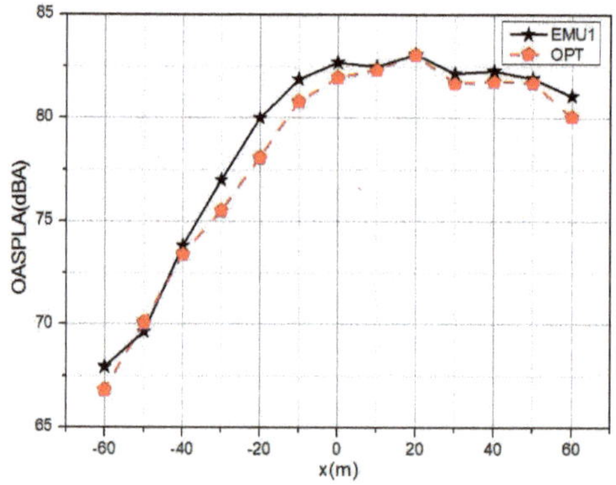

Figure 18. The Overall A-weighted Sound Pressure Levels (OASPLAs) of all the probes before and after optimization.

It can be seen that the OASPLAs of most probes for the optimal shape are reduced in a certain extent, compared to the prototype. Results reveal that the noise circumstance in the far field is improved after optimization.

In order to investigate on the influence of each design variable on the optimization objectives, it is necessary to perform the sensitivity analysis for the design variables. Sensitivity analysis is performed by studying the correlations between the design variables and the objectives. The correlation r between variable x and y takes the form:

$$r = \frac{N\sum xy - (\sum x)(\sum y)}{\sqrt{\left[N\sum x^2 - (\sum x)^2\right]\left[N\sum y^2 - (\sum y)^2\right]}} \tag{3}$$

where N is the number of pairs of values.

Figure 19 shows the student charts between the objectives and the design variables.

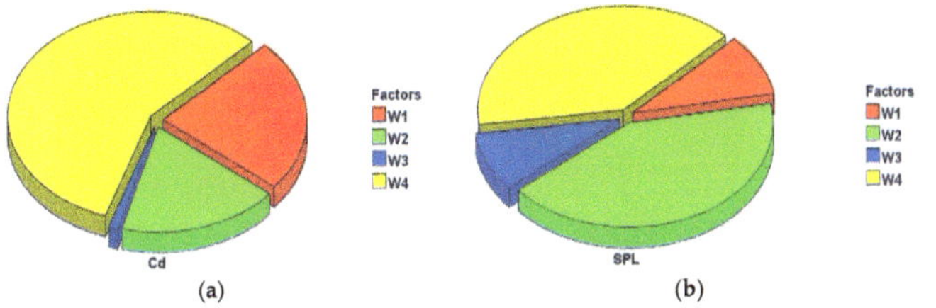

Figure 19. Student charts between the objectives and the design variables: (**a**) student chart between drag and four design variables; and (**b**) student chart between sound pressure level and four design variables.

As stated in Section 3, W_1 represents the width of the cab window, W_2 represents the height of the cab window, W_3 represents the height of the nose, and W_4 represents the width around the drainage. The design space is constrained under the real circumstance. Taking W_3 as an example, it should be not too much smaller or too much larger, or the connection between the hooks of adjacent streamlines could be affected. As seen above, W_4 has the largest influence on both the total drag and the aerodynamic noise. Besides, W_1 and W_2 own relatively large influence on the total drag and aerodynamic noise, respectively.

The three-dimensional relationship could also be established on the base of the final surrogate model. Figure 20a shows the relationship between aerodynamic noise and W_2 and W_4, while Figure 20b is the projection of aerodynamic noise on the W_2–W_4 plane. Strong nonlinearity could be observed. To facilitate analysis, we keep one of the design variables unchanged; the relationships of aerodynamic noise and the other design variable could be obtained, as shown in Figure 21.

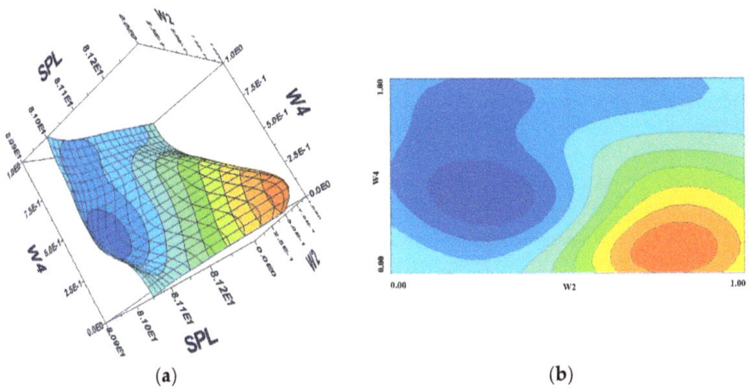

Figure 20. Relationship between aerodynamic noise and W_2 and W_4. (**a**) Surrogate model between aerodynamic noise and W_2 and W_4; and (**b**) projection of aerodynamic noise on the W_2–W_4 plane.

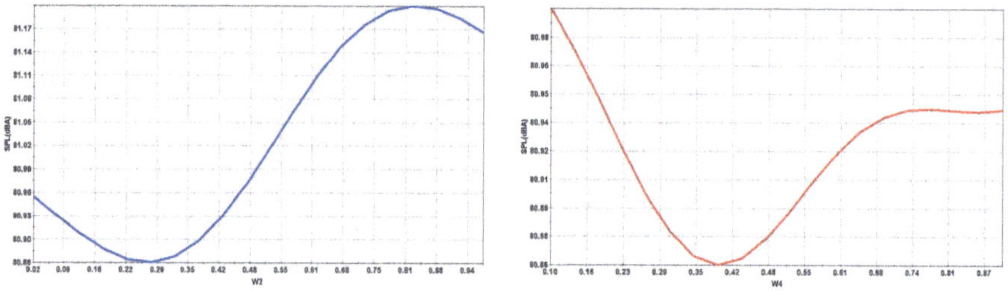

(**a**) (**b**)

Figure 21. The relationships of aerodynamic noise and the other design variable on the condition of one design variable unchanged: (**a**) SPL vs. W_2; and (**b**) SPL vs. W_4.

Nonlinearity could also be seen in above figures, and best performance of aerodynamic noise does not appear at minimal or maximal values of W_2 or W_4. For example, it can be deduced from above figures that the width of the drainage could not be too thin or too wide, or aerodynamic noise would grow apparently.

7. Conclusions

An aerodynamic noise optimization strategy for high-speed trains has been proposed in the present paper. In order to take all the acoustic probes into consideration, the equivalent continuous sound pressure level (SPL) has been taken as the objective for noise optimization. In order to improve the optimization efficiency, the local shape function method has been utilized for streamlined shape parameterization, which owns stronger generality and could set design variables directly on the focused region. Meanwhile, the surrogate based optimization approach is adopted, taking the Kriging model on the base of cross-validation method as the surrogate model. Once the Kriging model is built up, multi-objective optimization by genetic algorithm could be performed on this model to avoid the computational cost of CFD/CAA for each design point, which is the key for efficiency improvement.

Taking the drag of the whole train and the equivalent continuous SPL as the objectives, a multi-objective optimization of the streamlined shape of HSTs is performed. Results reveal that the aerodynamic performance has been improved apparently compared to the prototype. The drag of the whole train has been reduced by 7.1%, and the equivalent continuous SPL has been reduced by 1.26 dB. The OASPLAs of the probes for the optimal shape are mostly reduced except for very few of them, indicating that the acoustic circumstance in the far field has been effectively improved and it is feasible to perform aerodynamic noise optimization for high-speed trains with the use of the strategy in the present paper.

Acknowledgments: This work was supported by National Natural Science Foundation of China under 11302233, and Computing Facility for Computational Mechanics Institute of Mechanics, Chinese Academy of Sciences is gratefully acknowledged.

Author Contributions: Zhenxu Sun and Guowei Yang conceived the whole numerical simulations; Ye Zhang contributed the optimization tools; and Zhenxu Sun wrote the paper.

Conflicts of Interest: The authors declare no conflict of interest.

References

1. Kalivoda, M.; Danneskiold-samsoe, U.; Kruger, F. EURailNoise: A study of European priorities and strategies for railway noise abatement. *J. Sound Vib.* **2003**, *267*, 387–396. [CrossRef]
2. Talotte, C. Aerodynamic noise: A critical survey. *J. Sound Vib.* **2000**, *231*, 549–562. [CrossRef]
3. Talotte, C.; Gautier, P.E.; Thompson, D.J.; Hanson, C. Identification, modelling and reduction potential of railway noise sources: A critical survey. *J. Sound Vib.* **2003**, *267*, 447–468. [CrossRef]

4. Sun, Z.X.; Song, J.J.; An, Y.R. Numerical Simulation of Aerodynamic Noise Generated by High Speed Trains. *Eng. Appl. Comput. Fluid Mech.* **2012**, *6*, 173–185. [CrossRef]
5. Noger, C.; Patrat, J.C.; Peube, J. Aero-acoustical study of the TGV pantograph recess. *J. Sound Vib.* **2000**, *231*, 563–575. [CrossRef]
6. Sun, Z.X.; Guo, D.L.; Yao, S.B.; Yang, G.W.; Li, M.G. Identification and Suppression of Noise Sources around High Speed Trains. *Eng. Appl. Comput. Fluid Mech.* **2013**, *7*, 131–143. [CrossRef]
7. Ikeda, M.; Morikawa, T.; Manade, K. Development of low aerodynamic noise pantograph for high speed train. *Int. Congr. Noise Control Eng.* **1994**, *1*, 169–178.
8. Ikeda, M.; Suzuki, M.; Yoshida, K. Study on optimization of panhead shape possessing low noise and stable characteristics. *Q. Rep. RTRI* **2006**, *47*, 72–77. [CrossRef]
9. Iwamoto, K.; Higashi, A. Some consideration toward reducing aerodynamic noise on pantograph. *Jpn. Railw. Eng.* **1993**, *122*, 1–4.
10. Sueki, T.; Ikeda, M.; Takaishi, T. Aerodynamic noise reduction using porous materials and their application to high-speed pantographs. *Q. Rep. Railw. Tech. Res. Inst.* **2009**, *50*, 26–31. [CrossRef]
11. Lee, J.; Kim, J. Approximate optimization of high-speed train nose shape for reducing micropressure wave. *Ind. Appl.* **2008**, *35*, 79–87. [CrossRef]
12. Vytla, V.V.; Huang, P.G.; Penmetsa, R.C. Multi Objective Aerodynamic Shape Optimization of High Speed Train Nose Using Adaptive Surrogate Model. In Proceedings of the 28th AIAA Applied Aerodynamics Conference, Chicago, IL, USA, 28 June–1 July 2010.
13. Ku, Y.C.; Kwak, M.H.; Park, H.I.; Lee, D.H. Multi-Objective Optimization of High-Speed Train Nose Shape Using the Vehicle Modeling Function. In Proceedings of the 48th AIAA Aerospace Sciences Meeting, Orlando, FL, USA, 4–7 January 2010.
14. Yao, S.B.; Guo, D.L.; Sun, Z.X.; Yang, G.W.; Chen, D.W. Multi-objective optimization of the streamlined head of high-speed trains based on the Kriging model. *Sci. China Technol. Sci.* **2012**, *55*, 3494–3508. [CrossRef]
15. Suzuki, M.; Nakade, K. Multi-objective Design Optimization of High-speed Train Nose. *J. Mech. Syst. Transp. Logist.* **2013**, *6*, 54–64. [CrossRef]
16. Muñoz-Paniagua, J.; García, J.; Crespo, A. Genetically aerodynamic optimization of the nose shape of a high-speed train entering a tunnel. *J. Wind Eng. Ind. Aerodyn.* **2014**, *130*, 48–61. [CrossRef]
17. Batten, P.; Ribaldone, E.; Casella, M.; Chakravarthy, S. Towards a Generalized Non-Linear Acoustics Solver. In Proceedings of the 10th AIAA/CEAS Aeroacoustics Conferences, Manchester, UK, 10–12 May 2004.
18. Merci, B.; Vierendeels, J.; Langhe, C.D.; Dick, E. Development and Application of a New Cubic Low-Reynolds Eddy-Viscosity Turbulence Model. In Proceedings of the 31st AIAA Fluid Dynamics Conference & Exhibit, Anaheim, CA, USA, 11–14 June 2001.
19. Williams, J.E.F.; Hawkings, D.L. Sound generation by turbulence and surfaces in arbitrary motion. *Philos. Trans. R. Soc. Lond. Ser. A Math. Phys. Sci.* **1969**, *264*, 321–342. [CrossRef]
20. Casper, J.; Farassat, F. A new time domain formulation for broadband noise predictions. *Int. J. Aeroacoustics* **2002**, *1*, 207–240. [CrossRef]
21. Lee, I.; Sung, H.J. Characteristics of wall pressure fluctuations in separated and reattaching flows over a backward-facing step: Part 1. Time-mean statistics and cross-spectral analyses. *Exp. Fluids* **2001**, *30*, 262–272. [CrossRef]
22. Blazek, J. *Computational Fluid Dynamics Principles and Applications*; Elsevier Ltd.: Amsterdam, The Netherlands, 2005; pp. 16–17.
23. Menter, F.R. Zonal Two Equation K-W Turbulence Models for Aerodynamic Flows. In Proceedings of the 24th AIAA Fluid Dynamics Conference, Orlando, FL, USA, 6–9 July 1993.
24. Sederberg, T.W.; Parry, S.R. Free-Form deformation of solid geometric models. *SIGGRAPH Comput. Graph.* **1986**, *20*, 151–160. [CrossRef]
25. Piegl, L.; Tiller, W. The NURBS Book. In *Monographs in Visual Communication*; Springer: Berlin/Heidelberg, Germany, 1997.
26. Noel, C. The Origins of Kriging. *Math. Geol.* **1990**, *22*, 239–252.
27. Donald, R.J.; Matthias, S.; William, J.W. Efficient Global Optimization of Expensive Black-Box Functions. *J. Glob. Optim.* **1998**, *13*, 455–492.
28. Donald, R.; Jones, A. Taxonomy of Global Optimization Methods Based on Response Surfaces. *J. Glob. Optim.* **2001**, *21*, 345–383.

29. Gao, Y.H.; Wang, X.C. An effective warpage optimization method in injection molding based on Kriging model. *Int. J. Adv. Manuf. Technol.* **2008**, *37*, 953–960. [CrossRef]
30. Markatou, M.; Biswas, S. Analysis of variance of cross-validation estimators of the generalization error. *J. Mach. Learn. Res.* **2005**, *6*, 1127–1168.

© 2017 by the authors. Licensee MDPI, Basel, Switzerland. This article is an open access article distributed under the terms and conditions of the Creative Commons Attribution (CC BY) license (http://creativecommons.org/licenses/by/4.0/).

applied sciences

MDPI

Article

Vibro-Acoustic Numerical Analysis for the Chain Cover of a Car Engine

Enrico Armentani [1,*], Raffaele Sepe [1], Antonio Parente [2] and Mauro Pirelli [2]

[1] Department of Chemical, Materials and Production Engineering University of Naples Federico II, P.le V. Tecchio 80, 80125 Naples, Italy; raffsepe@unina.it
[2] Fiat Chrysler Automobiles (FCA) Powertrain S.p.A., via ex Aeroporto, 80038 Pomigliano D'Arco, Italy; antonio.parente2@fcagroup.com (A.P.); mauro.pirelli@fcagroup.com (M.P.)
* Correspondence: enrico.armentani@unina.it

Academic Editors: Roberto Citarella and Luigi Federico
Received: 20 January 2017; Accepted: 8 June 2017; Published: 12 June 2017

Abstract: In this work, a vibro-acoustic numerical and experimental analysis was carried out for the chain cover of a low powered four-cylinder four-stroke diesel engine, belonging to the FPT (FCA Power Train) family called SDE (Small Diesel Engine). By applying a methodology used in the acoustic optimization of new FPT engine components, firstly a finite element model (FEM) of the engine was defined, then a vibration analysis was performed for the whole engine (modal analysis), and finally a forced response analysis was developed for the only chain cover (separated from the overall engine). The boundary conditions applied to the chain cover were the accelerations experimentally measured by accelerometers located at the points of connection among chain cover, head cover, and crankcase. Subsequently, a boundary element (BE) model of the only chain cover was realized to determine the chain cover noise emission, starting from the previously calculated structural vibrations. The numerical vibro-acoustic outcomes were compared with those experimentally observed, obtaining a good correlation. All the information thus obtained allowed the identification of those critical areas, in terms of noise generation, in which to undertake necessary improvements.

Keywords: vibration analysis; acoustic analysis; FEM; BEM; NVH; modal analysis

---

1. Introduction

The new more stringent environmental laws in the automotive industry induced original equipment manufacturers (OEMs) and their suppliers and partners to address their efforts towards eco-friendly vehicle development, keeping low noise emissions and high passenger comfort levels. This is because—especially for premium cars—the interior acoustics play an important role for the drive comfort and drivers' subjective perception of quality, which in turn influences the customer's satisfaction and furthermore the purchase decision.

Currently, research and development is focusing strongly on multi-material applications for lightweight bodies by combining aluminum, magnesium, high-strength steels, Carbon Fiber Reinforced Polymer (CFRP), and organic sheets. Such materials will help to meet future requirements for lower weight, higher safety, and increased strength. However, lightweight materials are easily activated from a vibrational point of view (lightweight structures exhibit lower mass damping effects on vibrations), making them noisier. In order to design an improved lightweight vehicle body with suitable vibration and acoustics properties, the structural dynamics, vibrating behavior, sound transmission, and radiating performances must be examined in detail (e.g., some aspects of the plastic timing chain cover performances are presented in [1]). With reference to the powertrain, a reduction of noise emission can be obtained by optimizing cover structures (e.g., stiffening them or adding damping material).

Besides combustion noise, the noise generated by mechanical contacts of the moving and vibrating parts in an engine contributes to the overall engine noise. Depending on the engine operating condition, this mechanical noise may become the dominant part [2].

In order to improve the lifetime and reliability performances of the timing drive systems used in the automotive industry, the engines are often equipped with chain drives that become—with their vibrations—possible sources of disturbing noise. An early assessment of possible negative acoustical effects of the chain drive by the excitation of surrounding structures can significantly reduce the costs for prototyping and successive changes in the layout of the chain drive and engine.

The main inner excitation mechanism in the chain drive is the polygonal action, which means that a chain engaging on a sprocket forms a polygon rather than a circle due to the discretization of the sprocket teeth which couples with chain rollers. Since the polygonal effect occurs with the meshing frequency, the excited vibrations are basically narrow-banded and can finally be evaluated as an annoying whine-noise (the typical noise excited by polygonal chain drive action is named chain whine or meshing noise). The so-called polygon frequency depends on the number of teeth and the rotational speed of the sprocket. In general, the polygon frequency lies between some 100 and some 1000 Hz [3].

Commonly, only a few elastic structures (e.g., shafts) are considered in dynamic chain drive simulations. To consider the fully elastic behavior of arbitrary complex structures of chain drive components and the surrounding structure, a possible approach combines the multibody representation of the chain with an elastic Finite Element (FE) structure [4,5].

The main parameters that influence the system behavior and hence the acoustic performance of the model are:

- stiffness and damping of chain, guides, shafts/sprockets, bearings, and surrounding structure;
- inertias of chain, sprockets, plunger (tensioner), and tensioning guide;
- friction between chain–sprocket, chain–guides, chain link–sprocket;
- behavior of tensioner: damping (leakage, ventilation), and stiffness (oil, spring).

Another problem of the timing system equipped with chain is the elongation due to wear, responsible for an increase of noise due to impact and vibrations.

An integrated experimental and analytical modeling approach aimed at examining the impact noise characteristics of automotive-type timing chain drive systems was presented in [6], where a comparative study between various system parameters and operating conditions was provided.

An acoustic model relating the dynamic response of rollers and corresponding induced sound pressure levels is developed in [7], where the acceleration response of each chain roller for the noise level prediction of a chain drive under varying operating conditions was studied by a FEM (finite element method) approach.

Timing chain systems were modeled in [8] by a theoretical approach considering mass viscoelastic contact characteristics of each link or tooth separately.

A new roller chain sprocket tooth geometry for engine camshaft drives was studied in [9], having as a target the reduction of noise levels during meshing. Experimental tests were carried out to compare the noise levels of the asymmetrical sprocket tooth profile to that of a standard ISO sprocket tooth profile.

Experimental studies of timing chain vibration aimed at avoiding acoustically critical excitations were carried out in [10].

The modeling of the entire timing drive system containing a bushing chain drive, camshafts, and all connected single valve train was presented in [11], where the primary dynamics of the chain drive and the forces which are transferred to the engine's structure were assessed.

The torsional oscillations of engine camshafts induced by valve train loads were studied in [12]: they can cause significant tension fluctuation in the timing chain with corresponding increase in the chain transverse vibrations; this in turn causes an undesirable impact noise from the chain or sprocket meshing process.

In this work, the use of acoustic boundary elements and structural finite elements is aimed at the prediction of the radiation of structural noise from a side cover. A comparison is made between numerical results and experimental measurements, and a discussion is provided for the practical application of these modeling methods to side cover design. As a matter of fact, a computational method to allow reproducing the actual performance of a component for a car engine from a vibro-acoustic point of view was developed.

The vibro-acoustic numerical and experimental analysis was carried out for the chain cover of a low-powered four-cylinder four-stroke diesel engine (Figure 1a), belonging to the FPT (Fiat Power Train) family called SDE (Small Diesel Engine).

(a) (b)

Figure 1. (**a**) Fiat Power Train (FPT) Small Diesel Engine (SDE); (**b**) Finite element model (FEM) model of the engine.

By applying a methodology used in the acoustic optimization of new FPT engine components, firstly an FEM approach was applied for a modal analysis of the overall engine and also for a forced response analysis of the only chain cover. For the latter, the applied boundary conditions were the accelerations experimentally measured by accelerometers located at the points of connection among chain cover, head cover, and crankcase. Subsequently, a boundary element model of the only chain cover was realized to determine the chain cover noise emission, starting from the previously calculated vibrations [13]. The numerical vibro-acoustic outcomes were compared with those experimentally observed.

All the information thus obtained allowed the identification of those critical areas—in terms of noise generation—on which to undertake necessary improvements.

The modelling and computation were carried out with the support of the Hypermesh [14] and Abaqus [15] codes for the FEM modelling and modal analysis, and SIEMENS-LMS/Virtual Lab [16] for the BEM (boundary element method) modelling and vibro-acoustic analysis. In particular, Hypermesh code was used for the FEM mesh generation. The aforementioned software suites were selected being those in use in FPT at the time the work was performed.

2. Description of the Methodology

A vibro-acoustic analysis for a chain cover of a four-stroke four-cylinder diesel engine was performed through an FEM–BEM coupled approach. Such a hybrid approach—selected as an alternative to an FEM–FEM approach—is particularly advantageous when considering a free field acoustic emission, because with BEM it suffices to model the vibrating finite boundary, being null the contribution of integrals evaluated on the infinite boundary (Sommerfeld condition). Moreover, even for closed-domain acoustic emission problems, there can be computational advantages by coupling FEM and BEM approaches [17].

In [17], the reader can find detailed explanations about the way in which the interface between FEM and BEM approaches is realized (e.g., with reference to mapping of results from the FE-mesh to the BE-mesh with corresponding interpolation errors caused by such a data transfer process).

The vibration analysis is executed using FEM, whereas BEM is adopted for the simulation of the free field acoustic emission because it is more computationally efficient than FEM (BEM does not need to model the remote boundary at infinity as provided by the Sommerfeld radiation condition). A free field FE-based acoustic simulation would also be possible, but special FEM elements (infinite finite elements) would be necessary in this case.

The interface between the programs is realized by automatic data exchange tools available in the Virtual Lab environment following a theoretical approach detailed in [17] and synthetically reported in Section 4.

The BE approach combined with the acoustic transfer vector (ATV) technique in principle allows the acoustic emission to be evaluated in a shorter computational time when considering different engine regimes and various stiffening configurations.

Having to map a vibrational scenario from a refined FEM mesh to a coarse BEM mesh, some minor interpolation errors are introduced by the interfacing process.

Components showing a direct structural interaction with the chain cover were explicitly included in the FEM model: tappet cover, head, crankcase, sub-crankcase with bearing caps, and oil pan.

The FEM overall model (Figure 1b) consisted of 546,155 linear elements with a size of 6 mm and 507,978 nodes, whereas the chain cover was modelled by using quadratic elements with a size of 3 mm.

The overall FEM engine model is only adopted for a free-free modal analysis from which to extract the chain cover modal basis.

The modal analysis basis of the chain cover was extracted, leveraging on the results provided by the modal analysis of the overall engine. Afterwards, applying the vibrational excitations experimentally measured at the bolted connections between the chain cover and the engine block, a modal-based forced response analysis was performed for the chain cover.

The interface between chain cover and engine block was modelled using tie constraints everywhere except for the connection bolts where rigid elements (RBE2) are used. Equipment for the experimental vibrational measurements was available and consequently exploited to get a realistic assessment of the vibrational input to the chain cover.

The vibration numerical results were compared with corresponding experimental measurements.

The BEM model (Figure 2) was based on a discretization involving the component surface (outer "skin" of the component) and consisted of 1300 surface linear elements and 1385 nodes.

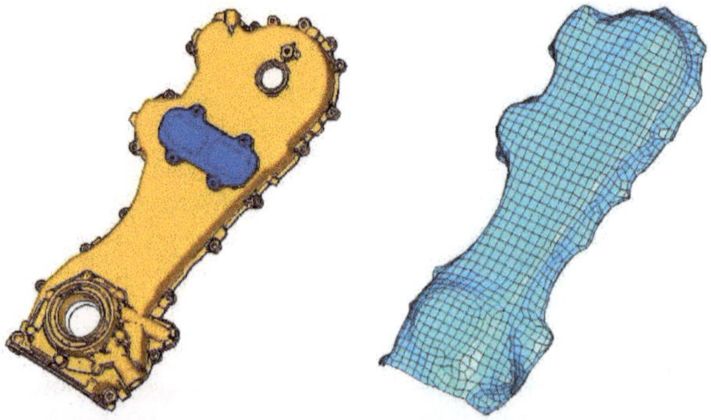

Figure 2. Boundary element method (BEM) model of the chain cover.

The objective of this work was the evaluation of noise emission from the sole chain cover: in this respect, the BEM discretization was only applied to the component under examination.

The environmental acoustic pressure distribution was calculated by the BEM model, and then compared with the related experimental acquisitions.

3. Acoustic Analysis

A BEM analysis was carried out to provide the acoustic emission of the chain cover. It is important to underline that the acoustic calculation evaluates the noise emission solely due to structurally transmitted excitation vibrations (structural born noise). The airborne noise and that emitted by other sources is not taken into account in any way.

The BEM analysis was performed, resorting to the indirect BEM [18] (quantities at the surface are jump of pressure and jump of normal velocity) with a variational approach for the numerical resolution [19,20].

The acoustic calculation is performed on a modal basis and involves different stages [17]:

- creation of a BEM mesh;
- calculation of the ATVs;
- introduction of experimentally-measured vibration velocities on the BEM mesh;
- calculation of the sound pressure and sound power.

As mentioned, the BEM mesh represents the mesh of the component surface.

The ATVs [21] are transfer functions allowing the calculation of the sound pressure in the chain cover surrounding environment, starting from the structure vibration.

The ATVs depend on the geometry and impedance of the vibrating surface, the microphones position, the frequency, and the physical properties of the domain in which the sound propagates (temperature and density).

The calculation of the ATVs is independent from the vibration result, and consequently, it does not depend on the loading conditions and damping values used in the structural calculation.

If the vibrating surface is divided into a discrete number of elements, the relationship between their normal velocities and the sound pressure in multiple domain points can be expressed as follows:

$$\{p(\omega)\} = [ATV(\omega)]\{v_n(\omega)\} \quad (1)$$

where $\{p(\omega)\}$ is the column vector containing the acoustic pressure in the different points of interest, $[ATV(\omega)]$ is the transfer vectors matrix, and $\{v_n(\omega)\}$ is the column vector containing the normal velocities of vibrating elements (as obtained by experimental measurements).

In particular, ATV is the output of a BEM problem solution, and represents the transfer function relating the vibration velocity in a node of the structure to the sound measured in a point of the acoustic domain.

The aforementioned computational processes represent a quite easy tool for the acoustic calculation. In fact, once the mesh of the component has been defined, it turns to the ATVs computation.

The ATVs' independence from the structural modal response entails a remarkable advantage in terms of computational times reduction, since any further modification to the component—which does not require a variation of its outer surface (acoustic emission side)—can be processed without the need for an ex-novo ATVs computation.

ATVs would only change if the external (emission side) chain cover design changed, whereas they would not be affected by structural changes on the remaining part of engine or changes in the excitations.

The main interest of this work stands in the assessment of chain cover emission in order to optimize its shape. The idea is to apply ribs on the chain cover internal side in order to stiffen it and correspondingly reduce the acoustic emission versus variable engine regimes.

The boundaries of the BE model enclose a domain; to avoid the presence of anomalous peaks in correspondence of the eigenfrequencies of such enclosed acoustic cavity, a specific impedance was assigned to the internal surface of the elements (and therefore to the material), equal to that of air—that is, 416.5 Rayl ($kg \cdot s^{-1} \cdot m^{-2}$).

Having placed the microphone in the chain cover near field, it is reasonable to assume that the contribution from parts other than the chain cover to the recorded sound pressure level turns out to be negligible.

4. Experimental Data

The vibrations transmitted by the engine running at 850 rpm were measured by placing 17 unidirectional accelerometers in correspondence with the attachment points of the chain cover to the rest of the engine (Figure 3), and placing a unidirectional accelerometer (for this kind of application, piezoelectric accelerometers perform pretty well) on top of the cover (Figure 4). Moreover, a microphone was placed in proximity of the cover (near-field microphone) (Figure 5). A microphone array was avoided because, due to its extension, it would be sensible to noise coming from parts other than the chain cover.

The standard ISO 3745 recommends the use of one microphone at the distribution side, one at the exhaust, one at the inlet, one for the timing of valves, and another one for gears. Being only interested in the (near-field) distribution emission, only the corresponding microphone was arranged.

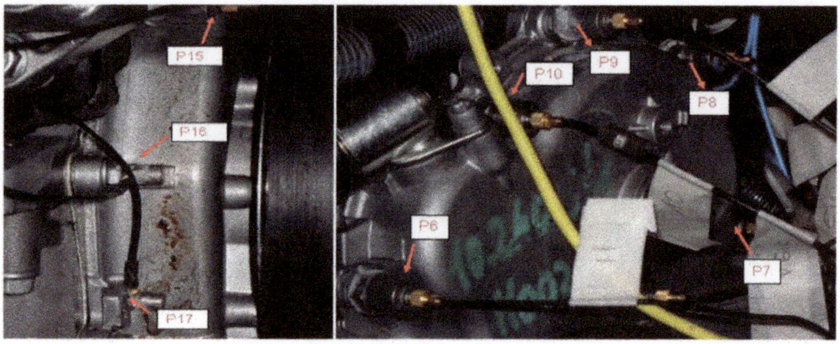

Figure 3. Accelerometers placed in proximity to the connecting pins between the chain cover and the rest of the engine.

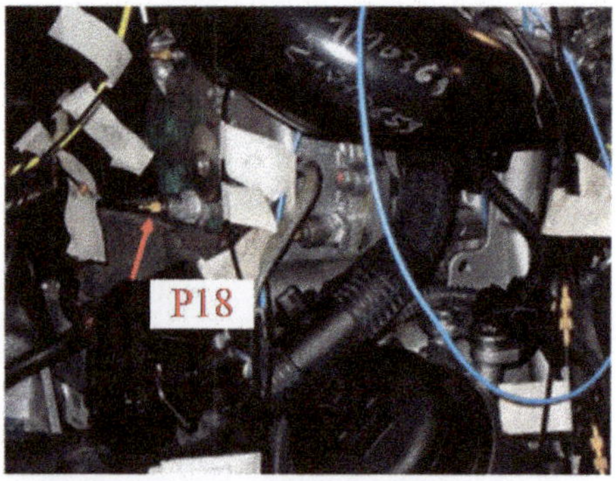

Figure 4. Accelerometer placed on the chain cover.

Figure 5. Near-field microphone.

The accelerometers placed on the fasteners detect the excitations acting on the cover in terms of local imposed accelerations; the latter are subsequently integrated by the code for vibro-acoustic assessment in order to obtain a profile of vibrational velocities (boundary conditions for the acoustic analysis). The accelerometer placed on the cover surface detects the vibration response of the chain cover, to be compared with the numerical vibration response. The near-field microphone—due to its short distance from the cover—essentially measures the sound pressure generated by the cover vibration phenomenon.

5. Vibro-Acoustic Validation

The experimental data allow the values of the excitation input for the BEM computational model to be obtained, to be introduced in the form of velocities at the attachment points of the chain cover (Figure 6).

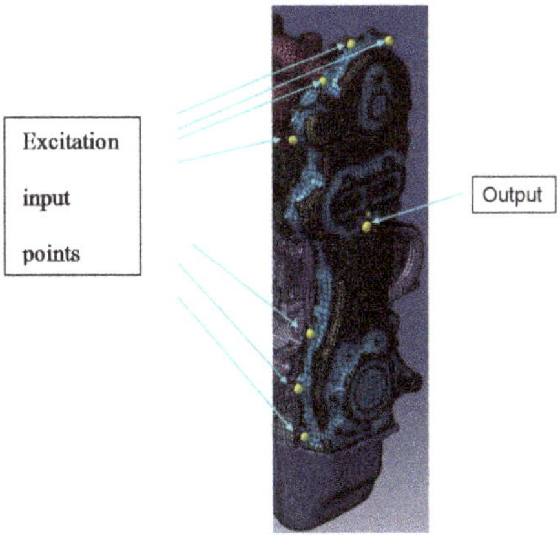

Figure 6. Vibro-acoustic model for the vibration analysis.

In the model for the numerical vibration analysis, 17 nodes in which the measured accelerations are converted into velocities were inserted at the attachment points between the cover and the engine. Moreover, in addition to the selected position for the accelerometer placed on the chain cover, also a node constituting the point in which the forced response is detected is taken into consideration (Figure 6).

The code for vibro-acoustic assessment allows the determination of the forced vibration response of the narrow-band model in the frequency range 1–4000 Hz (this is the frequency range of interest).

The relevant modes of the chain cover are in the range 1–3 kHz, and the excitation is mainly related to the engine regime that is generally lower than 4 kHz (e.g., the relevant combustion frequencies are much lower, being in a range lower than 200 Hz). Consequently, there is no interest to go further with the maximum frequency of considered eigenmodes. On the other hand, a further acoustic and structural mesh refinement would be needed to operate at higher frequencies.

It can be observed how the behavior of the numerical and experimental curves are consistent with each other (Figure 7).

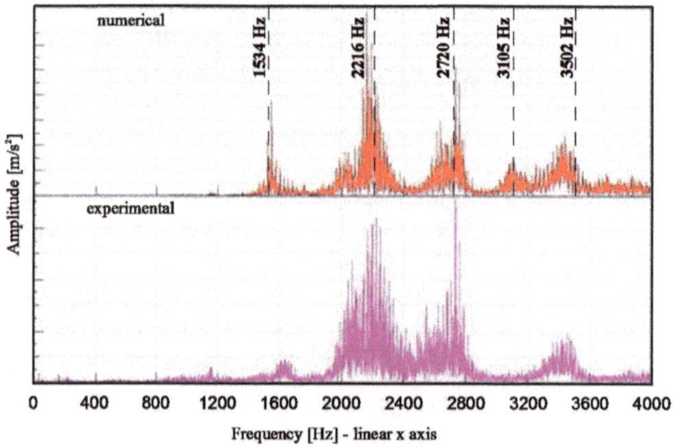

Figure 7. Narrow-band numerical (**up**) and experimental (**down**) comparison of the acceleration at the measurement point.

For the numerical acoustic analysis, a virtual microphone was placed in the position where the experimental outcomes are measured (Figure 8). The narrow-band sound pressure in the range 1–4000 Hz was first obtained, and then compared with data experimentally acquired, showing a satisfactory correlation (Figure 9).

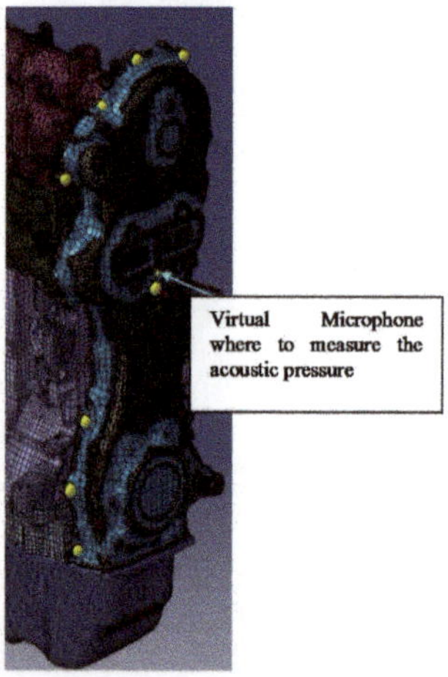

Figure 8. Vibro-acoustic model for the acoustic analysis.

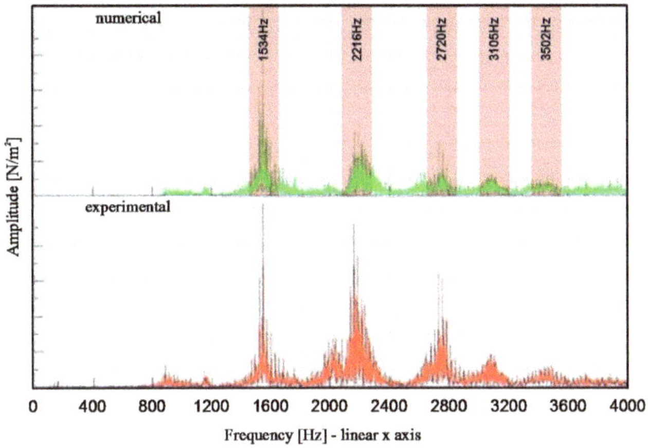

Figure 9. Narrow-band numerical (**up**) and experimental (**down**) comparison of the sound pressure magnitude.

The dominant frequency of 3200 Hz is missing in the numerical result: one possible reason could be related to the approximation inherent to the assumption of a uniform modal damping (3–5%) in the whole frequency range.

The results related to the sound pressure allowed the resonant frequencies to be identified, and therefore the corresponding resonant modes for which the noise emission of the chain cover is maximum (Figure 9).

Once the resonant frequencies and modes are known (Table 1), for each mode, the areas of the cover exhibiting higher emissions are visible (Figure 10).

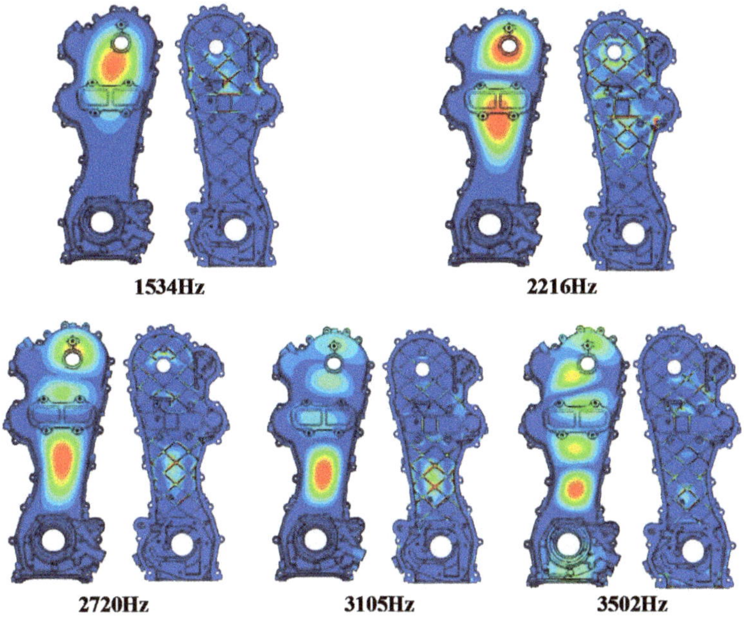

Figure 10. Chain cover eigenmodes.

Table 1. Critical modes.

Critical Modes	
MODE 1	1534 Hz
MODE 2	2216 Hz
MODE 3	2720 Hz
MODE 4	3105 Hz
MODE 5	3502 Hz

The obtained information can be used during the component design/development phase, in order to reduce its noise emissions.

6. Conclusions

In this work, a computational model for the vibro-acoustic analysis of the chain cover of a car diesel engine was developed.

The numerical results were compared with the experimental measurements, showing a good correlation.

The component noise emission was evaluated by isolating the highest-emission areas and correlating them with the local cover modes.

This method represents a valid tool to be used in the project design phase for the optimization of the acoustic transparency of the engine components, since it allows the implementation of appropriate changes without the need for prototypes, considerably reducing the development costs and times.

Author Contributions: A.P. and M.P. conceived, designed and performed the experiments; E.A. and R.S. analyzed the data and wrote the paper.

Conflicts of Interest: The authors declare that there is no conflict of interest regarding the publication of this paper.

References

1. Oh, K.; Han, W.; Jang, J.; Tho, Y.; Kim, H.H. *Design for Nvh Performance and Weight Reduction in Plastic Timing Chain Cover Application*; SAE Technical Paper 2014-01-1043; SAE International: Warrendale, PA, USA, 2014.
2. Priebsch, H.H.; Herbst, H.; Offner, G.; Sopouch, M. Numerical simulation and verification of mechanical noise generation in combustion engines. In Proceedings of the ISMA2002 Conference, Leuven, Belgium, 16–18 September 2002.
3. Howlett, M.; Ausserhofer, N.; Shoeffmann, W.; Truffinet, C.; Zurk, A. Chain versus belt—System comparison of future driving times. *Int. J. Automot. Eng.* **2016**, *7*, 135–141.
4. Siano, D.; Citarella, R. Elastic Multi Body Simulation of a Multi-Cylinder Engine. *Open Mech. Eng. J.* **2014**, *8*, 157–169. [CrossRef]
5. Armentani, E.; Sbarbati, F.; Perrella, M.; Citarella, R. Dynamic analysis of a car engine valve train system. *Int. J. Veh. Noise Vib.* **2016**, *12*, 229–240. [CrossRef]
6. Liu, S.P.; Dent, A.M.; Thornton, J.W.; Trethewey, K.W.; Wang, K.W.; Hayek, S.I.; Chen, F.H.K. *Experimental Evaluation of Automotive Timing Chain Drive Impact Noise*; SAE Technical Paper 951239; SAE International: Warrendale, PA, USA, 1995.
7. Zheng, H.; Wang, Y.Y.; Liu, G.R.; Lam, K.Y.; Quek, K.P.; Ito, T.; Noguchi, Y. Efficient modelling and prediction of meshing noise from chain drives. *J. Sound Vib.* **2001**, *245*, 133–150. [CrossRef]
8. Sopouch, M.; Hellinger, W.; Priebsch, H.H. Prediction of vibro acoustic excitation due to the timing chains of reciprocating engines. *Proc. Inst. Mech. Eng. K* **2003**, *217*, 225–240.
9. Young, J.D.; Marshek, K. Camshaft Roller Chain Drive with Reduced Meshing Impact Noise Levels. In Proceedings of the SAE Noise and Vibration Conference, Traverse City, MI, USA, 5–9 May 2003.
10. Calvo, J.A.; Diaz, V.; Roman, J.L.S.; Ramirez, M. Controlling the timing chain noise in diesel engines. *Int. J. Veh. Noise Vib.* **2006**, *2*, 75–90. [CrossRef]
11. Sopouch, M.; Hellinger, W.; Priebsch, H. *Simulation of Engine's Structure Borne Noise Excitation Due to the Timing Chain Drive*; SAE Technical Paper 2002-01-0451; SAE International: Warrendale, PA, USA, 2002.

12. Wang, K. *Vibration Analysis of Engine Timing Chain Drives with Camshaft Torsional Excitations*; SAE Technical Paper 911063; SAE International: Warrendale, PA, USA, 1991.
13. Armentani, E.; Trapani, R.; Citarella, R.; Parente, A.; Pirelli, M. FEM-BEM Numerical Procedure for Insertion Loss Assessment of an Engine Beauty Cover. *Open Mech. Eng. J.* **2013**, *7*, 27–34. [CrossRef]
14. Altair Engineering. *Hypermesh User Manual*; Altair Engineering: Troy, MI, USA, 2011.
15. Dassault Systèmes Simulia Corporation. *Abaqus Analysis User's Manual*; Version 6.12.1; SIMULIA: Providence, RI, USA, 2011.
16. SIEMENS-LMS Virtual Lab. *User Manual*; SIEMENS-LMS Virtual Lab: Plano, TX, USA, 2011.
17. Citarella, R.; Federico, L.; Cicatiello, A. Modal acoustic transfer vector approach in a FEM–BEM vibro-acoustic analysis. *Eng. Anal. Bound. Elem.* **2007**, *31*, 248–258. [CrossRef]
18. Citarella, R.; Landi, M. Acoustic analysis of an exhaust manifold by Indirect Boundary Element Method. *Open Mech. Eng. J.* **2011**, *5*, 138–151. [CrossRef]
19. Von Estorff, O.; Coyette, J.-P.; Migeot, J.-L. Governing formulations of the BEM in acoustics. In *Boundary Elements in Acoustics—Advances and Applications*; von Estorff, O., Ed.; WIT Press: Southampton, UK, 2000.
20. Migeot, J.-L.; Meerbergen, K.; Lecomte, C.; Coyette, J.-P. Practical implementation issues of acoustic BEM. In *Boundary Elements in Acoustics—Advances and Applications*; von Estorff, O., Ed.; WIT Press: Southampton, UK, 2000.
21. Gérard, F.; Tournour, M.; el Masri, N.; Cremers, L.; Felice, M.; Selmane, A. Acoustic Transfer Vectors for Numerical Modeling of Engine Noise. *Sound Vib. Mag.* **2002**, *36*, 20–25.

© 2017 by the authors. Licensee MDPI, Basel, Switzerland. This article is an open access article distributed under the terms and conditions of the Creative Commons Attribution (CC BY) license (http://creativecommons.org/licenses/by/4.0/).

applied sciences

MDPI

Article

Computational Vibroacoustics in Low- and Medium- Frequency Bands: Damping, ROM, and UQ Modeling

Roger Ohayon [1] and Christian Soize [2,*]

1 Structural Mechanics and Coupled Systems Laboratory, Conservatoire National des Arts et Metiers (CNAM), 2 rue Conte, 75003 Paris, France; roger.ohayon@cnam.fr
2 Laboratoire Modélisation et Simulation Multi-Echelle (MSME UMR 8208 CNRS), Université Paris-Est, 5 bd Descartes, 77454 Marne-la-Vallée, France
* Correspondence: christian.soize@univ-paris-est.fr; Tel.: +33-160-957-661

Academic Editors: Roberto Citarella and Luigi Federico
Received: 10 May 2017; Accepted: 3 June 2017; Published: 7 June 2017

Abstract: Within the framework of the state-of-the-art, this paper presents a summary of some common research works carried out by the authors concerning computational methods for the prediction of the responses in the frequency domain of general linear dissipative vibroacoustics (structural-acoustic) systems for liquid and gas in the low-frequency (LF) and medium-frequency (MF) domains, including uncertainty quantification (UQ) that plays an important role in the MF domain. The system under consideration consists of a deformable dissipative structure, coupled with an internal dissipative acoustic fluid including a wall acoustic impedance, and surrounded by an infinite acoustic fluid. The system is submitted to given internal and external acoustic sources and to prescribed mechanical forces. An efficient reduced-order computational model (ROM) is constructed using a finite element discretization (FEM) for the structure and the internal acoustic fluid. The external acoustic fluid is treated using a symmetric boundary element method (BEM) in the frequency domain. All the required modeling aspects required for the analysis in the MF domain have been introduced, in particular the frequency-dependent damping phenomena and model uncertainties. An industrial application to a complex computational vibroacoustic model of an automobile is presented.

Keywords: structural acoustics; vibroacoustic; uncertainty quantification; reduced-order model; medium frequency; low frequency; dissipative system; wall acoustic impedance; finite element discretization; boundary element method

1. Introduction

This paper presents a summary of some common research works carried out by the authors concerning computational methods for the prediction of the responses in the frequency domain of general linear dissipative vibroacoustics (structural-acoustic) systems for liquid and gas in the low- frequency (LF) and medium-frequency (MF) domains, including uncertainty quantification (UQ) that plays an important role in the MF domain. The contribution of this paper is the presentation of an efficient computational methodology adapted to large-scale computational vibroacoustic models which corresponds to a combination of established methods, and an application of this computational methodology is presented for an industrial vibroacoustic system. Considering all the aspects that are developed in this paper in order to present a complete strategy for the modeling and computational approaches—external acoustic fluid modeling, internal acoustic fluid including a dissipation term and frequency-dependent wall impedance, structure with frequency-dependent constitutive equation, reduced-order model for weak and strong coupling between the structure and the internal acoustic fluid, complete methodology for uncertainty quantification—a synthetic presentation has been adopted for readability, and we refer the reader to the given appropriate bibliography for the details. Nevertheless, since the model uncertainties induced by the modeling errors—which cannot be taken into account by using the usual probabilistic parametric approach of uncertain parameters or of other deterministic approaches of uncertainties—is relatively novel for the readers who are not familiarized with such a formulation, this part is more detailed in terms of equations.

More specifically, this paper is devoted to computational methods for the prediction of the frequency responses of linear dissipative vibroacoustic (structural-acoustic) systems in the low- and medium-frequency ranges. The vibroacoustic system consists of a deformable dissipative structure, coupled with a bounded internal dissipative acoustic fluid including a wall acoustic impedance, immersed in an unbounded acoustic fluid, and submitted to internal and external acoustic sources, as well as mechanical forces.

Computational vibroacoustic predictions play an important and increasing role in analyzing industrial complex systems, and many works have been published in this field.

The physics on which the computational vibroacoustics formulations are based can be found in numerous books, such as in [1–11].

For the spatial discretization of structures and bounded internal acoustic fluids, the computational vibroacoustics are generally based on the finite element method (FEM) (see for instance [12–15], and [16] for the corresponding isogeometric formulation).

For the spatial discretization of the unbounded external acoustic fluid, either finite element method or boundary element method (BEM) are used. Concerning the use of the local discretization of the unbounded external acoustic fluid, we refer the reader to [17–26], and for the BEM that is based on the finite element discretization of the boundary integral equation methods, let us cite [27–37]. Concerning the BEMs that are specifically devoted to the unbounded external acoustic fluid, we refer the reader to [38–48].

Reduced-order models (ROMs) are very attractive and efficient for analyzing large computational vibroacoustics models that have a significant number of design parameters—also called parametric high-dimensional computational models (HDMs)—for design and optimization, for constructing online models for active control, and for taking uncertainties into account. Particularly for parametric nonlinear HDMs, many approaches which can also be applied to parametric linear HDMs have been proposed, including hyper-reduced ROM, which guarantees feasibility [49–61]. With these methods, the parameter admissible space must be sampled at a few points using a greedy sampling algorithm (e.g., [62]), and a set of problems must be solved, yielding a set of parametric solution snapshots. This generally results from a proper orthogonal decomposition (POD), which are compressed using, for example, singular value decomposition (SVD) to construct a global reduced-order basis (ROB). For linear vibroacoustic HDMs, the most common choice for the parametric ROB consists of taking the modes of the different parts that constitute the vibroacoustic system (e.g., the elastic modes of the structure and the acoustic modes of the internal acoustic fluid)[41,63–74].

The design of a vibroacoustic system is used to manufacture a real system and to construct a nominal computational model with the methodologies listed above, and which will be presented in the next sections. In practice, the real system can exhibit variabilities in its responses due to fluctuations in the manufacturing process and due to small variations of the configuration around a nominal configuration associated with the design. The vibroacoustic computational model has parameters such as geometry, mechanical properties, and boundary conditions, which can be uncertain, inducing uncertainties in the computational model parameters. On the other hand, the modeling process induces some modeling errors defined as the model uncertainties. It is important to consider both the model-parameters uncertainties and the modeling uncertainties in order to improve the predictions. Two main types of approaches can be used to model uncertainties in the framework of the probability theory [75]. The first is the parametric probabilistic approach, which consists of modeling the uncertain parameters by random variables (e.g., [75–84]), but which does not have the capability to take modeling uncertainties into account. Two main methods can then be used to take them into account. The first is the output-prediction-error method, which requires experimental data [85,86] and uses the Bayesian method [87,88], but for which the computational model cannot be updated using experimental data, which constitutes a lock for robust design and optimization. The second one is the nonparametric probabilistic approach of modeling uncertainties induced by modeling errors proposed in [75,89–91], based on the maximum entropy principle [92] in the context of Information Theory [93] and on the random matrix theory [94], and extended for nonlinear dynamical systems in [95].

The outline of this paper is the following:

- Statement of the problem in the frequency domain.
- External inviscid acoustic fluid equations in the frequency domain and acoustic impedance boundary operator.
- Internal dissipative acoustic fluid equations in the frequency domain.
- Structure equations with frequency-dependent constitutive equation.
- Boundary value problem in terms of the structural displacement and the internal pressure field.
- Vibroacoustic computational model.
- Reduced-order vibroacoustic computational model.
- Uncertainty quantification for the vibroacoustic computational model.
- Experimental validation with a complex computational vibroacoustic model of an automobile.

2. Statement of the Problem in the Frequency Domain

The physical space $\mathbb{R}^3$ refers to a cartesian reference system, and the generic point of $\mathbb{R}^3$ is denoted by $\mathbf{x} = (x_1, x_2, x_3)$. For any function $f(\mathbf{x})$, the notation $f_{,j}$ designates the partial derivative with respect to x_j. The classical convention for summations over repeated Latin indices is used, but not over Greek indices. The vibration problem is formulated in the frequency domain. Therefore, the Fourier transform is introduced for various quantities involved. For instance, for the displacement field $(\mathbf{x}, t) \mapsto \mathbf{u}(\mathbf{x}, t)$, the simplified notation $(\mathbf{x}, \omega) \mapsto \mathbf{u}(\mathbf{x}, \omega) = \int_{-\infty}^{+\infty} e^{-i\omega t}\, \mathbf{u}(\mathbf{x}, t)\, dt$ is introduced and consists of using the same symbol for a quantity and its Fourier transform, in which the circular frequency ω (rad/s) is real. The vibroacoustic system is assumed to be in linear vibrations around a static equilibrium state taken as a natural state at rest.

Structure. In general, the structure of a complex vibroacoustic system is composed of a main part called the master structure that is accessible to conventional modeling including uncertainties modeling, and a secondary part called the fuzzy substructure related to the structural complexity and including for example many equipment units attached to the master structure. In the present paper, we will not consider fuzzy substructures, and concerning fuzzy structure theory, we refer the reader to Chapter 15 of [41] for a synthesis, and to [96] for an extension of the theory to the modeling of an uncertain complex vibroacoustic system with fuzzy interface. At equilibrium, the structure occupies the three-dimensional bounded domain Ω_S with a boundary $\partial\Omega_S$ that is made up of a part Γ_E that is the coupling interface between the structure and the external acoustic fluid, a part Γ that is a coupling interface between the structure and the internal acoustic fluid, and finally, a part Γ_Z that is another part of the coupling interface between the structure and the internal acoustic fluid with acoustic properties. The structure is assumed to be free (free-free structure). The outward unit normal to $\partial\Omega_S$ is denoted as $\mathbf{n}^S = (n_1^S, n_2^S, n_3^S)$ (see Figure 1). In Ω_S, the displacement field is denoted by $\mathbf{u}(\mathbf{x}, \omega) = (u_1(\mathbf{x}, \omega), u_2(\mathbf{x}, \omega), u_3(\mathbf{x}, \omega))$. A surface force field $\mathbf{G}(\mathbf{x}, \omega) = (G_1(\mathbf{x}, \omega), G_2(\mathbf{x}, \omega), G_3(\mathbf{x}, \omega))$ is given on $\partial\Omega_S$, and a body force field $\mathbf{g}(\mathbf{x}, \omega) = (g_1(\mathbf{x}, \omega), g_2(\mathbf{x}, \omega), g_3(\mathbf{x}, \omega))$ is given in Ω_S. The structure is a dissipative medium for which the frequency-dependent constitutive equation is detailed in Appendix A.

Internal dissipative acoustic fluid. Let Ω be the internal bounded domain filled with a dissipative acoustic fluid (gas or liquid) that is described in Section 4. The boundary $\partial\Omega$ of Ω is $\Gamma \cup \Gamma_Z$. The outward unit normal to $\partial\Omega$ is denoted by $\mathbf{n} = (n_1, n_2, n_3)$, and we have $\mathbf{n} = -\mathbf{n}^S$ on $\partial\Omega$ (see Figure 1).

The acoustic properties of boundary Γ_Z are modeled by a wall acoustic impedance $Z(\mathbf{x},\omega)$ satisfying the hypotheses defined in Section 4.2. In Ω, the pressure field is denoted by $p(\mathbf{x},\omega)$ and the velocity field by $\mathbf{v}(\mathbf{x},\omega)$. It is assumed that there is no Dirichlet boundary condition on any part of $\partial\Omega$. An acoustic source density $Q(\mathbf{x},\omega)$ is given inside Ω.

External inviscid acoustic fluid. The structure is surrounded by an external inviscid acoustic fluid (gas or liquid) that is detailed in Appendix B. The fluid occupies the infinite three-dimensional domain Ω_E whose boundary $\partial\Omega_E$ is Γ_E. The inward unit normal to $\partial\Gamma_E$ is $\mathbf{n}^S$, defined above (see Figure 1). In Ω_E, the pressure field is denoted by $p_E(\mathbf{x},\omega)$. There is no Dirichlet boundary condition on Γ_E. An acoustic source density $Q_E(\mathbf{x},\omega)$ is given in Ω_E. This acoustic source density induces a pressure field $p_{\text{given}}(\omega)$ on Γ_E (defined in Appendix B). For the sake of brevity, the case of an incident plane wave is not considered here, and the reader is referred to [41] for this case.

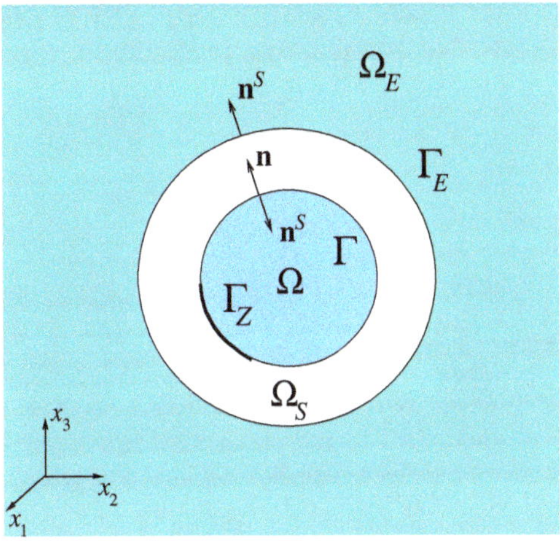

Figure 1. Geometry of the vibroacoustic system.

3. External Inviscid Acoustic Fluid Equations in the Frequency Domain and Acoustic Impedance Boundary Operator

Let ρ_E and c_E be the constant mass density and the constant speed of sound of the external acoustic fluid at equilibrium. Let $k = \omega/c_E$ be the wave number at frequency ω. The pressure is the solution of the classical exterior Neumann problem related to the Helmholtz equation with a source term [2–4,41],

$$\nabla^2 p_E + k^2\, p_E = -i\omega\, Q_E \quad \text{in} \quad \Omega_E\,, \tag{1}$$

$$\frac{\partial p_E}{\partial \mathbf{n}^S} = \omega^2\, \rho_E\, \mathbf{u}\cdot\mathbf{n}^S \quad \text{on} \quad \Gamma_E\,, \tag{2}$$

$$|\, p_E\,| = O(\frac{1}{R}) \quad , \quad \left|\frac{\partial p_E}{\partial R} + i\,k\,p_E\right| = O(\frac{1}{R^2})\,, \tag{3}$$

with $R = \|\mathbf{x}\| \to +\infty$, where $\partial/\partial R$ is the derivative in the radial direction and where $\mathbf{u}\cdot\mathbf{n}^S$ is the normal displacement field on Γ_E induced by the deformation of the structure. Equation (3) corresponds to the outward Sommerfeld radiation condition at infinity. In Appendix B, it is shown that the value $p_E|_{\Gamma_E}$ of the pressure field p_E on the external fluid–structure interface Γ_E is related to the given external pressure field $p_{\text{given}}|_{\Gamma_E}$ on Γ_E and to the normal displacement field $\mathbf{u}(\omega)\cdot\mathbf{n}^S$ on the external fluid–structure interface Γ_E by the equation

$$p_E|_{\Gamma_E}(\omega) = p_{\text{given}}|_{\Gamma_E}(\omega) + i\omega\, \mathbf{Z}_{\Gamma_E}(\omega)\{\mathbf{u}(\omega)\cdot\mathbf{n}^S\}\,, \tag{4}$$

in which $\mathbf{Z}_{\Gamma_E}(\omega)$ is the acoustic impedance boundary operator. It can be proven that for all given real ω and for all given $\mathbf{u}\cdot\mathbf{n}^S$ sufficiently regular on Γ_E, the boundary value problem defined by Equations (1)–(3) (e.g., [97,98]) admits a unique solution.

4. Internal Dissipative Acoustic Fluid Equations in the Frequency Domain

4.1. Internal Dissipative Acoustic Fluid Equations

The internal fluid is called a *dissipative acoustic fluid*, and is assumed to be homogeneous, compressible, dissipative, and at rest in the reference configuration. This dissipative acoustic fluid—for which the dissipation due to thermal conduction is neglected and for which the motions are assumed to be irrotational—is driven by the Helmholtz equation with an additional internal dissipative term. For additional details concerning dissipation in acoustic fluids, we refer the reader to [2,4,6,9]. Let ρ_0 be the mass density and let c_0 be the constant speed of sound in the acoustic fluid at equilibrium in the reference configuration Ω. The Helmholtz equation with a dissipative term and a source term Q is written as [41]

$$\frac{1}{\rho_0}\boldsymbol{\nabla}^2 p + i\omega\frac{\tau}{\rho_0}\boldsymbol{\nabla}^2 p + \frac{\omega^2}{\rho_0 c_0^2}\,p = -\frac{1}{\rho_0}(i\omega Q - \tau c_0^2\boldsymbol{\nabla}^2 Q) \quad \text{in} \quad \Omega\,, \tag{5}$$

in which τ is given by

$$\tau = \frac{1}{\rho_0 c_0^2}\left(\frac{4}{3}\eta + \zeta\right) > 0\,. \tag{6}$$

The constant η is the dynamic viscosity, $\nu = \eta/\rho_0$ is the kinematic viscosity, and ζ is the second viscosity which can depend on ω. Therefore, τ can depend on frequency ω. To simplify the notation, we write τ instead of $\tau(\omega)$. Taking $\tau = 0$ in Equation (5) yields the usual Helmholtz equation with an acoustic source for wave propagation in an inviscid acoustic fluid. It should be noted that the dissipation term is proportional to $\boldsymbol{\nabla}^2 p$ and not to p (what modifies the boundary condition defined after), and that the acoustic source term exhibits a $\boldsymbol{\nabla}^2 Q$ term.

4.2. Boundary Conditions

(i) Let $\mathbf{v}(\mathbf{x},\omega)$ be the velocity in the acoustic fluid. The interface condition on Γ for the inviscid dissipative acoustic fluid is written as $\mathbf{v}\cdot\mathbf{n} = i\omega\,\mathbf{u}\cdot\mathbf{n}$ on Γ. In terms of p, the Neumann boundary condition is written as

$$(1+i\omega\,\tau)\,\frac{\partial p}{\partial \mathbf{n}} = \omega^2\,\rho_0\,\mathbf{u}\cdot\mathbf{n} + \tau\,c_0^2\,\frac{\partial Q}{\partial \mathbf{n}} \quad \text{on} \quad \Gamma\,. \tag{7}$$

(ii) The wall acoustic impedance on Γ_Z is defined by the following constitutive equation:

$$p(\mathbf{x},\omega) = Z(\mathbf{x},\omega)\,\{\mathbf{v}(\mathbf{x},\omega)\cdot\mathbf{n} - i\omega\,\mathbf{u}(\mathbf{x},\omega)\cdot\mathbf{n}\}\,, \tag{8}$$

in which $Z(\mathbf{x},\omega)$ is the *wall acoustic impedance* defined for $\mathbf{x}\in\Gamma_Z$, with complex values. Wall acoustic impedance $Z(\mathbf{x},\omega)$ must satisfy appropriate conditions in order to ensure that the problem is correctly stated (see [41] for a general formulation). In terms of p, the Neumann boundary condition on Γ_Z is written as

$$(1+i\omega\,\tau)\,\frac{\partial p}{\partial \mathbf{n}} = \omega^2\,\rho_0\,\mathbf{u}\cdot\mathbf{n} - i\omega\rho_0\frac{p}{Z} + \tau\,c_0^2\,\frac{\partial Q}{\partial \mathbf{n}} \quad \text{on} \quad \Gamma_Z\,. \tag{9}$$

4.3. Case of a Free Surface for a Liquid

In a case of an acoustic liquid with a free surface Γ_0, neglecting gravity effects, the following Dirichlet condition is written as

$$p = 0 \quad \text{on} \quad \Gamma_0\,. \tag{10}$$

5. Structure Equations with Frequency-Dependent Constitutive Equation

Structure Equations in the Frequency Domain

The equation of the structure occupying domain Ω_S is written as

$$-\,\omega^2\,\rho_S\,u_i - \sigma_{ij,j}(\mathbf{u}) = g_i \quad \text{in} \quad \Omega_S\,, \tag{11}$$

in which $\rho_S(\mathbf{x})$ is the mass density of the structure. For a linear viscoelastic model, the symmetric stress tensor is written as

$$\sigma_{ij}(\mathbf{u}) = (a_{ijkh}(\omega) + i\omega\,b_{ijkh}(\omega))\,\varepsilon_{kh}(\mathbf{u})\,, \tag{12}$$

in which the symmetric strain tensor $\varepsilon_{kh}(\mathbf{u})$ is such that

$$\varepsilon_{kh}(\mathbf{u}) = \frac{1}{2}(u_{k,h}(\mathbf{x},\omega) + u_{h,k}(\mathbf{x},\omega))\,, \tag{13}$$

and where the tensors $a_{ijkh}(\omega)$ and $b_{ijkh}(\omega)$ depend on ω and are detailed in Appendix A for the LF and MF ranges.

On the fluid–structure external interface Γ_E, the boundary condition is such that

$$\sigma_{ij}(\mathbf{u})n_j^S = G_i - p_E|_{\Gamma_E}\,n_i^S \quad \text{on} \quad \Gamma_E\,. \tag{14}$$

Using Equation (4) yields

$$\sigma_{ij}(\mathbf{u})\,n_j^S = G_i - p_{\text{given}}|_{\Gamma_E}\,n_i^S - i\omega\,\mathbf{Z}_{\Gamma_E}(\omega)\{\mathbf{u}\cdot\mathbf{n}^S\}\,n_i^S \quad \text{on} \quad \Gamma_E\,. \tag{15}$$

Since $\mathbf{n}^S = -\mathbf{n}$, the boundary condition on $\Gamma \cup \Gamma_Z$ is written as

$$\sigma_{ij}(\mathbf{u})n_j^S = G_i + p\,n_i \quad \text{on} \quad \Gamma \cup \Gamma_Z\,, \tag{16}$$

in which p is the internal acoustic pressure field defined in Section 4.

6. Boundary Value Problem in Terms of $\{\mathbf{u}, p\}$

The boundary value problem in terms of $\{\mathbf{u}, p\}$ is written as follows. For all real ω and for given $\mathbf{G}(\omega)$, $\mathbf{g}(\omega)$, $p_{\text{given}}|_{\Gamma_E}(\omega)$, and $Q(\omega)$, find $\mathbf{u}(\omega)$ and $p(\omega)$, such that

$$-\,\omega^2\,\rho_S\,\mathbf{u} - \operatorname{div}\sigma(\mathbf{u}) = \mathbf{g} \quad \text{in} \quad \Omega_S\,, \tag{17}$$

$$\sigma(\mathbf{u})\,\mathbf{n}^S = \mathbf{G} - p_{\text{given}}|_{\Gamma_E}\,\mathbf{n}^S - i\omega\,\mathbf{Z}_{\Gamma_E}(\omega)\{\mathbf{u}\cdot\mathbf{n}^S\}\,\mathbf{n}^S \quad \text{on} \quad \Gamma_E\,, \tag{18}$$

$$\sigma(\mathbf{u})\,\mathbf{n}^S = \mathbf{G} + p\,\mathbf{n} \quad \text{on} \quad \Gamma \cup \Gamma_Z\,. \tag{19}$$

$$-\frac{\omega^2}{\rho_0 c_0^2}\,p - i\omega\frac{\tau}{\rho_0}\boldsymbol{\nabla}^2 p - \frac{1}{\rho_0}\,\boldsymbol{\nabla}^2 p = \frac{1}{\rho_0}(i\omega Q - \tau c_0^2\boldsymbol{\nabla}^2 Q) \quad \text{in} \quad \Omega\,. \tag{20}$$

$$(1 + i\omega\,\tau)\,\frac{\partial p}{\partial\mathbf{n}} = \omega^2\,\rho_0\,\mathbf{u}\cdot\mathbf{n} + \tau\,c_0^2\,\frac{\partial Q}{\partial\mathbf{n}} \quad \text{on} \quad \Gamma\,. \tag{21}$$

$$(1 + i\omega\,\tau)\,\frac{\partial p}{\partial\mathbf{n}} = \omega^2\,\rho_0\,\mathbf{u}\cdot\mathbf{n} - i\omega\rho_0\frac{p}{Z} + \tau\,c_0^2\,\frac{\partial Q}{\partial\mathbf{n}} \quad \text{on} \quad \Gamma_Z\,. \tag{22}$$

In the case of a free surface Γ_0 in the internal acoustic cavity (see Section 4.3), the boundary condition defined by Equation (10) must be added.

Remarks

- Equation (17) corresponds to the structure equation (see Equations (11)–(13)), in which $\{\text{div}\,\sigma(\mathbf{u})\}_i = \sigma_{ij,j}(\mathbf{u})$.
- Equations (18) and (19) are the boundary conditions for the structure (see Equations (15) and (16)).
- Equation (20) corresponds to the internal dissipative acoustic fluid equation (see Equation (5)).
- Equations (21) and (22) are the boundary conditions for the acoustic cavity (see Equations (7) and (9)).
- It is important to note that the external acoustic pressure field p_E has been eliminated as a function of $\mathbf{u}$ using the acoustic impedance boundary operator $\mathbf{Z}_{\Gamma_E}(\omega)$ (see Equation (4) of Section 3 and Appendix B), while the internal acoustic pressure field p is kept.

7. Vibroacoustic Computational Model

The finite element method is used to construct the spatial discretization of the variational formulation of the boundary value problem defined by Equations (17)–(22), with the additional boundary condition defined by Equation (10) in the case of a free surface for an internal liquid. We consider a finite element mesh of structure Ω_S and a finite element mesh of internal acoustic fluid Ω. It is assumed that the two finite element meshes are compatible on interface $\Gamma \cup \Gamma_Z$. The finite element mesh of surface Γ_E is the trace of the mesh of Ω_S.

7.1. Matrix Equation of the Computational Model

Let $\mathbb{U}(\omega)$ be the complex vector of the n_S degrees-of-freedom (DOFs), which are the values of $\mathbf{u}(\omega)$ at the nodes of the finite element mesh of domain Ω_S. For the internal acoustic fluid, let $\mathbb{P}(\omega)$ be the complex vectors of the n DOFs, which are the values of $p(\omega)$ at the nodes of the finite element mesh of domain Ω. The complex matrix equation of the computational model is written as

$$[\mathbb{A}_{\text{FSI}}(\omega)] \begin{bmatrix} \mathbb{U}(\omega) \\ \mathbb{P}(\omega) \end{bmatrix} = \begin{bmatrix} \mathbb{F}^S(\omega) \\ \mathbb{F}(\omega) \end{bmatrix} , \tag{23}$$

in which the complex matrix $[\mathbb{A}_{\text{FSI}}(\omega)]$ is defined by

$$\begin{bmatrix} [\mathbb{A}^S(\omega)] - \omega^2 [\mathbb{A}_{\text{BEM}}(\omega/c_E)] & [\mathbb{C}] \\ \omega^2\,[\mathbb{C}]^T & [\mathbb{A}(\omega)] + [\mathbb{A}^Z(\omega)] \end{bmatrix}, \tag{24}$$

in which $[\mathbb{C}]^T$ is the transposed matrix of $[\mathbb{C}]$. In Equation (24), the symmetric $(n_S \times n_S)$ complex matrix $[\mathbb{A}^S(\omega)]$ is defined by

$$[\mathbb{A}^S(\omega)] = -\omega^2 [\mathbb{M}^S] + i\omega\,[\mathbb{D}^S(\omega)] + [\mathbb{K}^S(\omega)] , \tag{25}$$

in which $[\mathbb{M}^S]$, $[\mathbb{D}^S(\omega)]$, and $[\mathbb{K}^S(\omega)]$ are symmetric $(n_S \times n_S)$ real matrices, which represent the mass matrix, the damping matrix, and the stiffness matrix of the structure. Matrix $[\mathbb{M}^S]$ is positive and invertible (positive definite), and matrices $[\mathbb{D}^S(\omega)]$ and $[\mathbb{K}^S(\omega)]$ are positive and not invertible (positive semidefinite) due to the presence of six rigid body motions (free-free structure). The symmetric $(n \times n)$ complex matrix $[\mathbb{A}(\omega)]$ is defined by

$$[\mathbb{A}(\omega)] = -\omega^2[\mathbb{M}] + i\omega\,[\mathbb{D}(\omega)] + [\mathbb{K}]\,, \tag{26}$$

in which $[\mathbb{M}]$, $[\mathbb{D}(\omega)]$, and $[\mathbb{K}]$ are symmetric $(n \times n)$ real matrices. Matrix $[\mathbb{M}]$ is positive and invertible, and matrices $[\mathbb{D}(\omega)]$ and $[\mathbb{K}]$ are positive and not invertible with rank $(n-1)$. In the case of a free surface Γ_0 in the internal acoustic cavity (see Section 4.3), the boundary condition defined by Equation (10) is added, and consequently the corresponding matrix $[\mathbb{K}]$ is positive definite, and is thus invertible. The internal fluid–structure coupling matrix $[\mathbb{C}]$ is related to the coupling between the structure and the internal fluid on the internal fluid–structure interface, and is a $(n_S \times n)$ real matrix that is only related to the values of $\mathbb{U}$ and $\mathbb{P}$ on the internal fluid–structure interface. The wall acoustic impedance matrix $[\mathbb{A}^Z(\omega)]$ is a symmetric $(n \times n)$ complex matrix depending on the wall acoustic impedance $Z(\mathbf{x},\omega)$ on Γ_Z, and which is only related to the values of $\mathbb{P}$ on boundary Γ_Z. The boundary element matrix $[\mathbb{A}_{\rm BEM}(\omega/c_{\rm E})]$—which depends on $\omega/c_{\rm E}$—is a symmetric $(n_S \times n_S)$ complex matrix that is only related to the values of $\mathbb{U}$ on the external fluid–structure interface Γ_E. This matrix is written as

$$[\mathbb{A}_{\rm BEM}(\omega/c_{\rm E})] = -\rho_{\rm E}\,[\mathbb{N}]^T\,[B_{\Gamma_E}(\omega/c_{\rm E})]\,[\mathbb{N}]\,, \tag{27}$$

in which $[B_{\Gamma_E}(\omega/c_{\rm E})]$ is the full symmetric $(n_E \times n_E)$ complex matrix defined in Appendix B, and where $[\mathbb{N}]$ is a sparse $(n_E \times n_S)$ real matrix related to the finite element discretization.

7.2. *Construction of the Matrices of the Computational Model*

The expressions of the real or complex bilinear forms whose discretization allows the corresponding real or complex matrices to be constructed are given hereinafter. For such a construction, we consider the fields $(p, \eta, \mathbf{u})$ and the corresponding fields $(\delta p, \delta\eta, \delta\mathbf{u})$ as test functions that are real (and not complex).

7.2.1. Matrices Related to the Equations of the Structure

- Symmetric real mass matrix $[\mathbb{M}^S]$ is positive definite, and corresponds to $\int_{\Omega_S} \rho_S\, \mathbf{u} \cdot \delta\mathbf{u}\, d\mathbf{x}$.
- Symmetric real damping matrix $[\mathbb{D}^S(\omega)]$ is positive semidefinite with a kernel of dimension 6, and corresponds to $\int_{\Omega_S} b_{ijkh}(\omega)\, \varepsilon_{kh}(\mathbf{u})\, \varepsilon_{ij}(\delta\mathbf{u})\, d\mathbf{x}$.
- Symmetric real stiffness matrix $[\mathbb{K}^S(\omega)]$ is positive semidefinite with a kernel of dimension 6, and corresponds to $\int_{\Omega_S} a_{ijkh}(\omega)\, \varepsilon_{kh}(\mathbf{u})\, \varepsilon_{ij}(\delta\mathbf{u})\, d\mathbf{x}$.

7.2.2. Matrices Related to the Equations of the Internal Acoustic Fluid

- Symmetric real matrix $[\mathbb{K}]$ is positive semidefinite with a kernel of dimension 1, and corresponds to $\frac{1}{\rho_0}\int_\Omega \boldsymbol{\nabla} p \cdot \boldsymbol{\nabla}\delta p\, d\mathbf{x}$. In the case of a free surface Γ_0 in the internal acoustic cavity (see Section 4.3), the boundary condition defined by Equation (10) is added, and consequently, the corresponding matrix $[\mathbb{K}]$ is positive definite.
- From Equation (20), it can be deduced that the symmetric real matrix $[\mathbb{D}(\omega)] = \tau(\omega)\,[\mathbb{K}]$ is positive semidefinite with a kernel of dimension 1, in which $\tau(\omega)$ is defined by Equation (6).
- Symmetric real matrix $[\mathbb{M}]$ is positive definite, and corresponds to $\frac{1}{\rho_0 c_0^2}\int_\Omega p\, \delta p\, d\mathbf{x}$.
- Symmetric complex matrix $[\mathbb{A}^Z(\omega)]$ comes from $\int_{\Gamma_Z} \frac{1}{Z(\omega)} p\, \delta p\, ds$, in which $Z(\mathbf{x},\omega)$ is a complex-valued function and where ds is the elementary surface area.

7.2.3. Matrices Related to the Coupling Terms

- Rectangular real matrix $[\mathbb{C}]$ corresponds to $-\int_{\Gamma\cup\Gamma_Z} p\, \mathbf{n} \cdot \delta\mathbf{u}\, ds$.

7.2.4. Vector of Mechanical and Acoustical Excitations

- Complex vector $\mathbb{F}^S(\omega)$ of external forces corresponds to $\int_{\Gamma_E}(\mathbf{G} - p_{\text{given}}|_{\Gamma_E}\,\mathbf{n}^S)\cdot\delta\mathbf{u}\,ds + \int_{\Omega_S}\mathbf{g}\cdot\delta\mathbf{u}\,d\mathbf{x}$.
- Complex vector $\mathbb{F}(\omega)$ of internal acoustic sources corresponds to $\frac{1}{\rho_0}\int_\Omega(i\omega\,Q\,\delta p + \tau c_0^2\nabla Q\,\cdot\nabla\delta p)\,d\mathbf{x}$.

Remark concerning the construction of the solution. The vibroacoustic computational model without model uncertainties (deterministic equations) can be numerically solved ω by ω, but for large-scale computational models, the numerical cost can be relatively high. This is why reduced-order models are generally introduced. In addition, if the nonparametric probabilistic approach is used to take the uncertainties induced by modeling errors into account, then a reduced-order model must be introduced. The methodology proposed for constructing a reduced-order model is presented in the next section.

8. Reduced-Order Vibroacoustic Computational Model

One possible method for constructing a reduced-order model would consist of choosing as a reduced-order basis the elastoacoustic modes (coupled vibroacoustic modes) of an associated conservative system that must be defined and which will be directly computed. The direct computation of these elastoacoustic modes can be very expensive for solving the eigenvalue problem of large-scale computational models, even if advanced algorithms based on Krylov methods are used. However, with such an approach, the nature of the couplings between the subsystems through the knowledge of the eigenfrequencies and the mode shapes of each subsystem can be difficult to analyze. In addition, the use of the elastoacoustic modes does not allow anymore for adapting the level of uncertainties as a function of the matrices of the coupled system (inertial forces, damping forces, stiffness forces in the structure, and equivalent forces in the internal dissipative acoustic fluid, internal fluid–structure coupling forces). This aspect is particularly important for the analyses in the medium-frequency range. Certainly, knowledge of the elastoacoustic modes can also be very useful for analyzing the consequences of the couplings. Their computation can be considerably decreased by using a reduced-order model such as the one proposed hereinafter, which avoids the expensive direct computation. Finally, if the direct computation of the elastoacoustic modes can be very useful for the cases for which there is a strong effect of the internal acoustic fluid on the structural modes (e.g., the case of a strong coupling such as a structure coupled with an internal liquid or of a lightweight structure coupled with and internal gas), the use of the reduced-order model presented in Section 8 is also very efficient (as explained in Section 8.1.2), because the structural modes are computed taking into account the quasistatic effect of the internal fluid (the added mass effect). Finally, in order to simplify the presentation of Section 8—which is devoted to the construction of the reduced-order model—we have considered the case for which the internal acoustic fluid is a gas and the case for which the internal acoustic fluid is a liquid in order to illustrate the two important cases of a weak coupling and of a strong coupling. Clearly, these two cases cover many coupling situations, such as a lightweight structure coupled with an internal gas for which a strong coupling can appear; in such a case, the formulation presented in Section 8.1.2 must be used. The construction of the reduced-order computational model requires appropriate projection bases to be introduced, as explained below.

- *Projection basis for the structure.*
  - If the internal acoustic fluid is a gas, the projection basis can be chosen as the undamped elastic structural modes of the structure in vacuo for which the constitutive equation corresponds to elastic materials (see Equation (A5)), and consequently, the stiffness matrix has to be constructed for $\omega = 0$.
  - If the internal acoustic fluid is a liquid (with or without free surface), for the structure, the projection basis can constructed as for a gas but by taking into account the effects of liquid's added mass.

- *Projection basis for the internal acoustic fluid.*
  - We consider the undamped acoustic modes of an internal acoustic cavity with fixed boundary (and rigid wall) and without wall acoustic impedance. Two cases must be considered: one for which the internal pressure varies with a variation of the volume of the cavity (a cavity with a sealed wall called a closed cavity), and the other one for which the internal pressure does not vary with the variation of the volume of the cavity (a cavity with a non-sealed wall, called an almost-closed cavity).

8.1. Computation of the Projection Basis for the Structure

8.1.1. Case of a Weak Coupling of the Structure with the Internal Acoustic Fluid

The undamped elastic structural modes of structure Ω_S in vacuo are computed with the constitutive equation corresponding to elastic materials. Setting $\lambda^S = \omega^2$, we have to solve the $(n_S \times n_S)$ generalized symmetric real eigenvalue problem,

$$[\mathbb{K}^S(0)]\,\mathbb{U} = \lambda^S\,[\mathbb{M}^S]\,\mathbb{U}\,. \tag{28}$$

This generalized eigenvalue problem admits a zero eigenvalue with multiplicity 6 (corresponding to the six rigid body motions), and admits an increasing sequence of $(n_S - 6)$ strictly positive eigenvalues (corresponding to the elastic structural modes),

$$0 < \lambda_1^S \leq \ldots \leq \lambda_\alpha^S \leq \ldots\,. \tag{29}$$

Let $\mathbb{U}_1, \ldots, \mathbb{U}_\alpha, \ldots$ be the eigenvectors (the elastic structural modes) associated with $\lambda_1^S, \ldots, \lambda_\alpha^S, \ldots$. Let $0 < N_S \leq n_S - 6$. We introduce the $(n_S \times N_S)$ real matrix of the N_S elastic structural modes $\mathbb{U}_\alpha$ associated with the first N_S strictly positive eigenvalues,

$$[\,\mathcal{U}\,] = [\,\mathbb{U}_1 \ldots \mathbb{U}_\alpha \ldots \mathbb{U}_{N_S}\,]\,. \tag{30}$$

We have the orthogonality properties,

$$[\,\mathcal{U}\,]^T\,[\mathbb{M}^S]\,[\,\mathcal{U}\,] = [M^S]\,, \tag{31}$$

$$[\,\mathcal{U}\,]^T\,[\mathbb{K}^S(0)]\,[\,\mathcal{U}\,] = [K^S(0)]\,, \tag{32}$$

in which $[M^S]$ is a diagonal matrix of positive real numbers and where $[K^S(0)]$ is the diagonal matrix of the eigenvalues such that $[K^S(0)]_{\alpha\beta} = \lambda_\alpha^S\,\delta_{\alpha\beta}$ (the eigenfrequencies are $\omega_\alpha^S = \sqrt{\lambda_\alpha^S}$).

8.1.2. Case of a Strong Coupling of the Structure with the Internal Acoustic Fluid

For the computation of the appropriate basis of the structure, the practical numerical procedure is given hereinafter. Equation (28) is replaced by

$$[\mathbb{K}^S(0)]\,\mathbb{U} = \lambda^S\,([\mathbb{M}^S] + [\mathbb{M}^A])\,\mathbb{U}\,, \tag{33}$$

in which $[\mathbb{M}^A]$ is the positive symmetric $(n_S \times n_S)$ real matrix (called the added mass matrix), which corresponds to the quasi-static effect of the internal acoustic fluid on the structure. It should be noted that the non-zero elements of matrix $[\mathbb{M}^A]$ are only related to the DOFs of the internal fluid–structure interface. This generalized eigenvalue problem admits a zero eigenvalue with multiplicity 6 (corresponding to the six rigid body motions) and admits an increasing sequence of $(n_S - 6)$ strictly positive eigenvalues,

$$0 < \lambda_1^S \leq \ldots \leq \lambda_\alpha^S \leq \ldots\,. \tag{34}$$

The eigenvectors $\mathbb{U}_1, \ldots, \mathbb{U}_\alpha, \ldots$ associated with $\lambda_1^S, \ldots, \lambda_\alpha^S, \ldots$ constitute a basis for the elastic structure. For $0 < N_S \leq n_S - 6$, we introduce the $(n_S \times N_S)$ real matrix of the N_S basis vectors $\mathbb{U}_\alpha$ associated with the first N_S strictly positive eigenvalues,

$$[\mathcal{U}] = [\,\mathbb{U}_1 \ldots \mathbb{U}_\alpha \ldots \mathbb{U}_{N_S}\,]\,. \tag{35}$$

We have the orthogonality properties,

$$[\mathcal{U}]^T\,([\mathbb{M}^S] + [\mathbb{M}^A])\,[\mathcal{U}] = [M^{SA}]\,, \tag{36}$$

$$[\mathcal{U}]^T\,[\mathbb{K}^S(0)]\,[\mathcal{U}] = [K^S(0)]\,, \tag{37}$$

in which $[M^{SA}]$ is a diagonal matrix of positive real numbers such that $[M^{SA}]_{\alpha\beta} = \mu_\alpha^S\,\delta_{\alpha\beta}$, and where $[K^S(0)]$ is the diagonal matrix of the eigenvalues such that $[K^S(0)]_{\alpha\beta} = \mu_\alpha^S\,\lambda_\alpha^S\,\delta_{\alpha\beta}$. It should be noted that $\omega_\alpha^S = \sqrt{\lambda_\alpha^S}$ is not an eigenfrequency of the structure in vacuo, but is an eigenfrequency of the structure with an added mass effect. For a closed acoustic cavity and for an internal acoustic liquid with a free surface, the construction of matrix $[\mathbb{M}^A]$ is given hereinafter.

Construction of added mass matrix $[\mathbb{M}^A]$ *for a closed acoustic cavity.* For a closed acoustic cavity, the deformations of the interface can induce a volume variation of the cavity. The positive symmetric $(n_S \times n_S)$ real matrix $[\mathbb{M}^A]$ is written as

$$[\mathbb{M}^A] = [\mathbb{C}]\,[\mathbb{S}]\,[\mathbb{C}]^T\,, \tag{38}$$

in which $[\mathbb{S}]$ is the positive symmetric $(n \times n)$ real matrix which is constructed in solving the following linear matrix equation

$$[\mathbb{K}]\,[\mathbb{S}] = [I_n]\,. \tag{39}$$

Under the constraint matrix equation

$$[\mathbb{B}]^T\,[\mathbb{S}] = [\,0\,]\,, \tag{40}$$

in which $[I_n]$ is the $(n \times n)$ identity matrix, $[\,0\,]$ is the $(1 \times n)$ zero matrix, and where $[\mathbb{B}]$ is the $(n \times 1)$ real matrix constructed by

$$[\mathbb{B}] = \rho_0\,c_0^2\,[\mathbb{M}]\,[\mathbb{1}]\,, \tag{41}$$

where $[\mathbb{1}]$ is the $(n \times 1)$ matrix such that $[\mathbb{1}]_{j1} = 1$ for all $j = 1, \ldots, n$. It should be noted that the numerical construction of matrix $[\mathbb{M}^A]$ can be viewed as the result of a Schur complement calculation with the constraint defined by Equation (40).

Construction of added mass matrix $[\mathbb{M}^A]$ *for an internal liquid with a free surface.* For an internal liquid with a free surface on which $p = 0$ (see Equation (10)), the introduced added mass corresponds to an incompressible fluid. The positive symmetric $(n_S \times n_S)$ real matrix $[\mathbb{M}^A]$ is written as

$$[\mathbb{M}^A] = [\mathbb{C}]\,[\mathbb{S}]\,[\mathbb{C}]^T\,, \tag{42}$$

in which $[\mathbb{S}]$ is the positive symmetric $(n \times n)$ real matrix which is constructed in solving the following linear matrix equation

$$[\mathbb{K}]\,[\mathbb{S}] = [I_n]\,, \tag{43}$$

in which $[\mathbb{K}]$ is invertible due to the free surface Γ_0 in the internal acoustic cavity. It should be noted that the numerical construction of matrix $[\mathbb{M}^A]$ can be viewed as the result of a Schur complement calculation.

8.2. Computation of the Projection Basis for the Internal Acoustic Fluid

8.2.1. Case of a Gas or a Liquid without Free Surface

The undamped acoustic modes of a closed (sealed wall) or an almost closed (non-sealed wall) acoustic cavity Ω are computed. Setting $\lambda = \omega^2$, we have the $(n \times n)$ generalized symmetric real eigenvalue problem

$$[\mathbb{K}]\,\mathbb{P} = \lambda\,[\mathbb{M}]\,\mathbb{P}\,. \tag{44}$$

This generalized eigenvalue problem admits a zero eigenvalue with multiplicity 1, denoted as λ_0 (corresponding to constant eigenvector denoted as $\mathbb{P}_0$) and admits an increasing sequence of $(n-1)$ strictly positive eigenvalues (corresponding to the acoustic modes),

$$0 < \lambda_1 \leq \ldots \leq \lambda_{n-1}\,. \tag{45}$$

Let $\mathbb{P}_1, \ldots, \mathbb{P}_{n-1}$ be the eigenvectors (the acoustic modes) associated with $\lambda_1, \ldots, \lambda_{n-1}$.

- *Closed (sealed wall) acoustic cavity.* Let be $0 < N \leq n$. We introduce the $(n \times N)$ real matrix of the constant eigenvector $\mathbb{P}_0$ and of the $(N-1)$ acoustic modes associated with the first $(N-1)$ strictly positive eigenvalues,

$$[\,\mathcal{P}\,] = [\,\mathbb{P}_0\,\mathbb{P}_1\,\ldots\,\mathbb{P}_{N-1}\,]\,. \tag{46}$$

- *Almost-closed (non-sealed wall) acoustic cavity.* Let be $0 < N \leq n-1$. We introduce the $(n \times N)$ real matrix of the N acoustic modes associated with the first N strictly positive eigenvalues,

$$[\,\mathcal{P}\,] = [\,\mathbb{P}_1\,\ldots\,\mathbb{P}_N\,]\,. \tag{47}$$

We have the orthogonality properties,

$$[\,\mathcal{P}\,]^T\,[\mathbb{M}]\,[\,\mathcal{P}\,] = [M]\,, \tag{48}$$

$$[\,\mathcal{P}\,]^T\,[\mathbb{K}]\,[\,\mathcal{P}\,] = [K]\,, \tag{49}$$

in which $[M]$ is a diagonal matrix of positive real numbers and where $[K]$ is the diagonal matrix of the eigenvalues such that $[K]_{\alpha\beta} = \lambda_\alpha\,\delta_{\alpha\beta}$ (for non-zero eigenvalue, the eigenfrequencies are $\omega_\alpha = \sqrt{\lambda_\alpha}$).

8.2.2. Case of a Liquid with a Free Surface

In such a case, there is a free surface Γ_0 in the internal acoustic cavity (see Section 4.3) for which the boundary condition defined by Equation (10) is added, and consequently the corresponding matrix $[\mathbb{K}]$ is positive definite. The undamped acoustic modes of the acoustic cavity Ω are computed by solving the $(n \times n)$ generalized symmetric real eigenvalue problem

$$[\mathbb{K}]\,\mathbb{P} = \lambda\,[\mathbb{M}]\,\mathbb{P}\,, \tag{50}$$

that admits an increasing sequence of n strictly positive eigenvalues (corresponding to the acoustic modes), $0 < \lambda_1 \leq \ldots \leq \lambda_n$, associated with the eigenvectors $\mathbb{P}_1, \ldots, \mathbb{P}_n$ (the acoustic modes). Let $0 < N \leq n$. We introduce the $(n \times N)$ real matrix of the N acoustic modes associated with the first N positive eigenvalues,

$$[\,\mathcal{P}\,] = [\,\mathbb{P}_1\,\ldots\,\mathbb{P}_N\,]\,. \tag{51}$$

Matrix $[\,\mathcal{P}\,]$ satisfies the orthogonality properties defined by Equations (48) and (49).

8.3. Construction of the Reduced-Order Computational Model

Using the projection basis $[\mathcal{U}]$ defined by Equation (35) for the structure, and the projection basis $[\mathcal{P}]$ defined by Equations (46), (47), or (51) for the acoustic fluid, the projection of Equation (23) yields the reduced-order computational model of order $N_S \ll n_S$ and $N \ll n$, which is written as

$$\mathbb{U}(\omega) = [\mathcal{U}]\, \mathbf{q}^S(\omega)\,, \tag{52}$$

$$\mathbb{P}(\omega) = [\mathcal{P}]\, \mathbf{q}(\omega)\,. \tag{53}$$

The complex vectors $\mathbf{q}^S(\omega)$ and $\mathbf{q}(\omega)$ of dimension N_S and N are the solution of the following equation

$$[A_{\rm FSI}(\omega)] \begin{bmatrix} \mathbf{q}^S(\omega) \\ \mathbf{q}(\omega) \end{bmatrix} = \begin{bmatrix} \mathbf{f}^S(\omega) \\ \mathbf{f}(\omega) \end{bmatrix} \quad , \tag{54}$$

in which the complex matrix $[A_{\rm FSI}(\omega)]$ is defined by

$$\begin{bmatrix} [A^S(\omega)] - \omega^2 [A_{\rm BEM}(\omega/c_{\rm E})] & [C] \\ \omega^2 [C]^T & [A(\omega)] + [A^Z(\omega)] \end{bmatrix} . \tag{55}$$

In Equation (55), the symmetric $(N_S \times N_S)$ complex matrix $[A^S(\omega)]$ is defined by

$$[A^S(\omega)] = -\omega^2 [M^S] + i\omega\, [D^S(\omega)] + [K^S(\omega)]\,, \tag{56}$$

in which $[M^S]$, $[D^S(\omega)]$, and $[K^S(\omega)]$ are positive-definite symmetric $(N_S \times N_S)$ real matrices such that

$$[D^S(\omega)] = [\mathcal{U}]^T\, [\mathbb{D}^S(\omega)]\, [\mathcal{U}] \quad , \quad [K^S(\omega)] = [\mathcal{U}]^T\, [\mathbb{K}^S(\omega)]\, [\mathcal{U}]\,. \tag{57}$$

The symmetric $(N \times N)$ complex matrix $[A(\omega)]$ is defined by

$$[A(\omega)] = -\omega^2 [M] + i\omega\, [D(\omega)] + [K]\,, \tag{58}$$

in which $[M]$, $[D(\omega)]$, and $[K]$ are symmetric $(N \times N)$ real matrices. For a closed (sealed wall) acoustic cavity, matrix $[K]$ is positive and not invertible with rank $N-1$, while for an almost-closed (non-sealed wall) acoustic cavity or for a liquid with a free surface (see Section 4.3), matrix $[K]$ is positive and invertible. Matrix $[M]$ is positive and invertible. The diagonal $(N \times N)$ real matrix $[D(\omega)]$ is written as

$$[D(\omega)] = \tau(\omega)\, [K]\,, \tag{59}$$

in which $\tau(\omega)$ is defined by Equation (6). The $(N_S \times N)$ real matrix $[C]$ is written as

$$[C] = [\mathcal{U}]^T\, [\mathbb{C}]\, [\mathcal{P}]\,. \tag{60}$$

The symmetric $(N \times N)$ complex matrix $[A^Z(\omega)]$ is such that

$$[A^Z(\omega)] = [\mathcal{P}]^T\, [\mathbb{A}^Z(\omega)]\, [\mathcal{P}]\,, \tag{61}$$

and finally, the symmetric $(N_S \times N_S)$ complex matrix $[A_{\rm BEM}(\omega/c_{\rm E})]$ is given by

$$[A_{\rm BEM}(\omega/c_{\rm E})] = [\mathcal{U}]^T\, [\mathbb{A}_{\rm BEM}(\omega/c_{\rm E})]\, [\mathcal{U}]\,. \tag{62}$$

The given forces are written as

$$\mathbf{f}^S(\omega) = [\mathcal{U}]^T\, \mathbb{F}^S(\omega) \quad , \quad \mathbf{f}(\omega) = [\mathcal{P}]^T\, \mathbb{F}(\omega)\,. \tag{63}$$

9. Uncertainty Quantification for the Vibroacoustic Computational Model

For the reasons given in Section 1, the nonparametric probabilistic approach of uncertainties is proposed for performing uncertainty quantification for the vibroacoustic computational model. It is recalled that such an approach is a way of modeling the uncertainties induced by the modeling errors that cannot be taken into account with the usual approaches that only consider the uncertainties on the physical parameters of the computational model. This nonparametric approach allows for separately but globally describing the model uncertainties that exist for each matrix appearing in the vibroacoustic computational model. There are only a small number of hyperparameters in the nonparametric probabilistic approach, these hyperparameters being the parameters of the probability distributions of the random matrices. This small number of hyperparameters allows for carrying out the experimental identification of the probabilistic model, which can easily be performed by solving a statistical inverse problem as explained in the application presented in Section 10. This kind of approach yields robust predictions with respect to model uncertainties that can effectively be quantified, for instance, by constructing the confidence region of the quantities of interest.

In this section, we summarize the fundamental concepts and the construction related to the nonparametric probabilistic approach of both computational model-parameters uncertainties and modeling uncertainties in computational vibroacoustics, detailed in [75,84]. The method presented has recently been implemented in MSC-Nastran^{TM} [99].

The methodology is applied to the reduced-order vibroacoustic computational model defined by Equations (52)–(58). In order to simplify the presentation, it is assumed that there are no uncertainties in the boundary element matrix $[A_{\text{BEM}}(\omega/c_{\text{E}})]$ or in the wall acoustic impedance matrix $[A^Z(\omega)]$. For fixed values of N_S and N, the stochastic reduced-order computational vibroacoustic model of order N_S and N is written as

$$\mathbf{U}(\omega) = [\mathcal{U}]\,\mathbf{Q}^S(\omega) \quad , \quad \mathbf{P}(\omega) = [\mathcal{P}]\,\mathbf{Q}(\omega)\,, \tag{64}$$

in which, for all real ω, the complex random vectors $\mathbf{Q}^S(\omega)$ and $\mathbf{Q}(\omega)$ of dimension N_S and N are the solution of the following random equation

$$[\mathbf{A}_{\text{FSI}}(\omega)] \begin{bmatrix} \mathbf{Q}^S(\omega) \\ \mathbf{Q}(\omega) \end{bmatrix} = \begin{bmatrix} \mathbf{f}^S(\omega) \\ \mathbf{f}(\omega) \end{bmatrix} \quad , \tag{65}$$

and where the complex random matrix $[\mathbf{A}_{\text{FSI}}(\omega)]$ is written as

$$\begin{bmatrix} [\mathbf{A}^S(\omega)] - \omega^2[A_{\text{BEM}}(\omega/c_{\text{E}})] & [\mathbf{C}] \\ \omega^2\,[\mathbf{C}]^T & [\mathbf{A}(\omega)] + [A^Z(\omega)] \end{bmatrix} . \tag{66}$$

The symmetric $(N_S \times N_S)$ complex random matrix $[\mathbf{A}^S(\omega)]$ is defined by

$$[\mathbf{A}^S(\omega)] = -\omega^2[\mathbf{M}^S] + i\omega\,[\mathbf{D}^S(\omega)] + [\mathbf{K}^S(\omega)]\,, \tag{67}$$

in which the probability distributions of the positive-definite symmetric $(N_S \times N_S)$ real random matrices $[\mathbf{M}^S]$, $[\mathbf{D}^S(\omega)]$, and $[\mathbf{K}^S(\omega)]$ are constructed in Sections 9.2 and 9.3. The symmetric $(N \times N)$ complex random matrix $[\mathbf{A}(\omega)]$ is written as

$$[\mathbf{A}(\omega)] = -\omega^2\,[\mathbf{M}] + i\omega\,[\mathbf{D}(\omega)] + [\mathbf{K}]\,, \tag{68}$$

in which $[\mathbf{M}]$, $[\mathbf{D}(\omega)]$ and $[\mathbf{K}]$ are symmetric $(N \times N)$ real random matrices. Random matrix $[\mathbf{M}]$ is positive definite. The diagonal $(N \times N)$ real random matrix $[\mathbf{D}(\omega)]$ is written as

$$[\mathbf{D}(\omega)] = \tau(\omega)\,[\mathbf{K}]\,, \tag{69}$$

in which $\tau(\omega)$ is deterministic and is defined by Equation (6). For a closed (sealed wall) acoustic cavity, random matrix $[\mathbf{K}]$ is positive and not invertible with rank $N-1$, while for an almost-closed (non-sealed wall) acoustic cavity or for a liquid with a free surface (see Section 4.3), random matrix $[\mathbf{K}]$ is positive definite. The probability distributions of random matrices $[\mathbf{M}]$, $[\mathbf{K}]$, and of the $(N_S \times N)$ real random matrix $[\mathbf{C}]$ are constructed in Sections 9.4–9.6.

9.1. Preliminary Results for the Stochastic Modeling of the Random Matrices for the Stochastic Reduced-Order Computational Vibroacoustic Model

In the framework of the nonparametric probabilistic approach of uncertainties, the probability distributions and the generators of independent realizations of such random matrices are constructed using random matrix theory [89,94] and the maximum entropy principle [92,100] from Information Theory [93], in which Shannon introduced the notion of entropy as a measure of the level of uncertainties for a probability distribution. For instance, if $p_X(x)$ is a probability density function on a real random variable X, the entropy $\mathcal{E}(p_X)$ of p_X is defined by $\mathcal{E}(p_X) = -\int_{-\infty}^{+\infty} p_X(x)\,\log(p_X(x))\,dx$. The maximum entropy principle consists of maximizing the entropy (i.e., maximizing the uncertainties), under the constraints defined by the available information. Consequently, it is important to define the algebraic properties of the random matrices for which the probability distributions have to be constructed. Let E be the mathematical expectation. For instance, for real-valued random variable X, we have $E\{X\} = \int_{-\infty}^{+\infty} x\,p_X(x)\,dx$ and $\mathcal{E}(p_X) = -E\{\log(p_X(X))\}$. In order to construct the probability distributions of the random matrices introduced before, we need to define a basic ensemble of random matrices.

It is known that a real Gaussian random variable can take negative values. Consequently, the Gaussian orthogonal ensemble (GOE) of random matrices [94]—which is the generalization for the matrix case of the Gaussian random variable—cannot be used when a positiveness property of the random matrix is required [75]. Therefore, new ensembles of random matrices are required to implement the nonparametric probabilistic approach of uncertainties.

9.1.1. Ensemble SG_0^+ of Random Matrices

Let $\mathbb{M}_0^+$ be the set of all the positive-definite symmetric $(m \times m)$ real matrices. Below, we summarize the construction [75,89,90] of an ensemble—denoted by SG_0^+—of random matrices with values in $\mathbb{M}_0^+$. An element of SG_0^+ is a positive-definite random matrix denoted by $[\mathbf{G}_0]$.

Definition of the available information for constructing ensemble SG_0^+. For a random matrix $[\mathbf{G}_0]$ belonging to SG_0^+, the available information consists of the mean value which is given and equal to the identity matrix, and an integrability condition that has to be imposed in order to ensure the decreasing of the probability density function around the origin,

$$E\{[\mathbf{G}_0]\} = [I_m]\,,\; E\{\log(\det[\mathbf{G}_0])\} = \chi\,, \tag{70}$$

in which $|\chi|$ is finite and where $[I_m]$ is the $(m \times m)$ identity matrix.

Probability density function of a matrix $[\mathbf{G}_0] \in SG_0^+$. The probability density function, $[G] \mapsto p_{[\mathbf{G}_0]}([G])$, defined on $\mathbb{M}_0^+$, of a random matrix $[\mathbf{G}_0]$ satisfies the usual normalization condition,

$$\int_{\mathbb{M}_0^+} p_{[\mathbf{G}_0]}([G])\,\widetilde{d}G = 1\,, \tag{71}$$

where the volume element $\widetilde{d}G$ is written as $\widetilde{d}G = 2^{m(m-1)/4}\,\Pi_{1\leq j\leq k\leq m}\, dG_{jk}$. Let δ be the positive real number defined by

$$\delta = \left\{ \frac{1}{m} E\{\|\,[\mathbf{G}_0] - [I_m]\,\|_F^2\} \right\}^{1/2}\,, \tag{72}$$

which will allow the dispersion of the probability model of random matrix $[\mathbf{G}_0]$ to be controlled, and where $\|\,\mathcal{M}\,\|_F$ is the Frobenius matrix norm of the matrix $[\mathcal{M}]$ such that $\|\,\mathcal{M}\,\|_F^2 = \mathrm{tr}\{[\mathcal{M}]^T[\mathcal{M}]\}$. For δ such that $0 < \delta < (m+1)^{1/2}(m+5)^{-1/2}$, the use of the maximum entropy principle under the constraints defined by Equations (70) and (71) yields for all $[G]$ in $\mathbb{M}_0^+$,

$$p_{[\mathbf{G}_0]}([G]) = c_0\big(\det{[G]}\big)^{c_1}\exp\{-c_2\,\mathrm{tr}[G]\}\,, \tag{73}$$

in which c_0 is the positive constant of normalization and where $c_1 = (m+1)(1-\delta^2)/(2\delta^2)$ and $c_2 = (m+1)/(2\delta^2)$.

Generator of independent realizations of a random matrix $[\mathbf{G}_0]$ **in** SG_0^+. The generator of independent realizations (which is required to solve the random equations with the Monte Carlo method) is constructed using the following algebraic representation of any random matrix $[\mathbf{G}_0]$ that belongs to SG_0^+,

$$[\mathbf{G}_0] = [\mathbf{L}]^T\,[\mathbf{L}]\,, \tag{74}$$

in which $[\mathbf{L}]$ is an upper triangular $(m \times m)$ random matrix such that:

- the family of the random entries $\{[\mathbf{L}]_{jj'}, j \leq j'\}$ are independent random variables;
- for $j < j'$, the real-valued random variable $[\mathbf{L}]_{jj'}$ is written as $[\mathbf{L}]_{jj'} = \sigma_m U_{jj'}$ in which $\sigma_m = \delta(m+1)^{-1/2}$ and where $U_{jj'}$ is a real-valued Gaussian random variable with zero mean and variance equal to 1;
- for $j = j'$, the positive-valued random variable $[\mathbf{L}]_{jj}$ is written as $[\mathbf{L}]_{jj} = \sigma_m\sqrt{2V_j}$ in which V_j is a positive-valued Gamma random variable with probability density function $\Gamma(a_j, 1)$ in which $a_j = \frac{m+1}{2\delta^2} + \frac{1-j}{2}$.

9.1.2. Ensemble $\mathrm{SG}_\varepsilon^+$ of Random Matrices

Let $0 \leq \varepsilon \ll 1$ be a positive number (for instance, ε can be chosen as 10^{-6}). We then define the ensemble $\mathrm{SG}_\varepsilon^+$ of all the random matrices such that

$$[\mathbf{G}] = \frac{1}{1+\varepsilon}\{[\mathbf{G}_0] + \varepsilon\,[I_m]\}\,, \tag{75}$$

in which $[\mathbf{G}_0]$ belongs to SG_0^+.

9.1.3. Cases of Several Random Matrices

It can be proven [75,91] that if there are several random matrices for which there is no available information concerning their statistical dependencies, then the use of the maximum entropy principle yields that the best model which maximizes the entropy (the uncertainties) is a stochastic model for which all these random matrices are independent.

9.2. Stochastic Modeling of Random Matrix $[\mathbf{M}^S]$

Since there is no available information concerning the statistical dependency of $[\mathbf{M}^S]$ with the other random matrices of the problem, then random matrix $[\mathbf{M}^S]$ is independent of all the other random matrices. The deterministic matrix $[M^S]$ is positive definite, and can consequently be written as

$[M^S] = [L_{M^S}]^T [L_{M^S}]$, in which $[L_{M^S}]$ is an upper triangular real matrix. Using the nonparametric probabilistic approach of uncertainties, the stochastic model of the positive-definite symmetric random matrix $[\mathbf{M}^S]$ is then defined by

$$[\mathbf{M}^S] = [L_{M^S}]^T [\mathbf{G}_{M^S}] [L_{M^S}], \tag{76}$$

where $[\mathbf{G}_{M^S}]$ is a $(N_S \times N_S)$ random matrix belonging to ensemble $\mathrm{SG}_\varepsilon^+$ defined in Section 9.1.2 and whose probability distribution and generator of independent realizations depend only on dimension N_S and on the dispersion parameter δ_{M^S}.

9.3. Stochastic Modeling of the Family $\{[\mathbf{D}^S(\omega)], [\mathbf{K}^S(\omega)]\}_\omega$ of Random Matrices Indexed by ω

Since there is no available information concerning the statistical dependency of the family $\{[\mathbf{D}^S(\omega)], [\mathbf{K}^S(\omega)]\}_\omega$ of random matrices with the other random matrices of the problem, then the family $\{[\mathbf{D}^S(\omega)], [\mathbf{K}^S(\omega)]\}_\omega$ are independent of all the other random matrices. However, we will see below that the families of random matrices $\{[\mathbf{D}^S(\omega)]\}_\omega$ and $\{[\mathbf{K}^S(\omega)]\}_\omega$ will be statistically dependent. For stochastic modeling of $\{[\mathbf{D}^S(\omega)]\}_\omega$ and $\{[\mathbf{K}^S(\omega)]\}_\omega$ related to the linear viscoelastic structure, we propose to use the extension presented in [101], which is based on the Hilbert transform [102] in the frequency domain to express the causality properties (similarly to the transforms used in Section A.2 of Appendix A). The nonparametric probabilistic approach of uncertainties then consists of modeling the positive-definite symmetric $(N_S \times N_S)$ real matrices $[D^S(\omega)]$ and $[K^S(\omega)]$ by random matrices $[\mathbf{D}^S(\omega)]$ and $[\mathbf{K}^S(\omega)]$ such that

$$E\{[\mathbf{D}^S(\omega)]\} = [D^S(\omega)] \quad , \quad E\{[\mathbf{K}^S(\omega)]\} = [K^S(\omega)], \tag{77}$$

$$[\mathbf{D}^S(-\omega)] = [\mathbf{D}(^S\omega)] \quad , \quad [\mathbf{K}^S(-\omega)] = [\mathbf{K}^S(\omega)]. \tag{78}$$

(i) For $\omega \geq 0$, the construction of the stochastic model of the family $\{[\mathbf{D}^S(\omega)], [\mathbf{K}^S(\omega)]\}_{\omega\geq 0}$ of random matrices is carried out as follows:

- Constructing the family $\{[\mathbf{D}^S(\omega)]\}_{\omega\geq 0}$ of random matrices such that

$$[\mathbf{D}^S(\omega)] = [L_{D^S}(\omega)]^T [\mathbf{G}_{D^S}] [L_{D^S}(\omega)], \tag{79}$$

  where $[L_{D^S}(\omega)]$ is such that

$$[D^S(\omega)] = [L_{D^S}(\omega)]^T [L_{D^S}(\omega)], \tag{80}$$

  and where $[\mathbf{G}_{D^S}]$ is a $(N_S \times N_S)$ random matrix belonging to ensemble $\mathrm{SG}_\varepsilon^+$, defined in Section 9.1.2. Its probability distribution and its generator of independent realizations depend only on dimension N_S and on the dispersion parameter δ_{D^S} that allows the level of uncertainties to be controlled.
- Constructing the family $\{[\widehat{\mathbf{N}}^R(\omega)]\}_{\omega\geq 0}$ of random matrices using the equation

$$[\widehat{\mathbf{N}}^R(\omega)] = \frac{2}{\pi}\mathrm{p.v}\int_0^{+\infty} \frac{\omega'^2}{\omega^2 - \omega'^2}[\mathbf{D}^S(\omega')]\, d\omega', \tag{81}$$

  (in which p.v means the Cauchy principal value that is defined in Equation (83)) or equivalently, using the two following equations that are useful for computation:

$$\text{for} \quad \omega = 0 \quad , \quad [\widehat{\mathbf{N}}^R(0)] \quad = \quad -\frac{2}{\pi}\int_0^{+\infty} [\mathbf{D}^S(\omega)]\, d\omega\,, \tag{82}$$

$$\begin{aligned} \text{for} \quad \omega > 0 \quad , \quad [\widehat{\mathbf{N}}^R(\omega)] \quad &= \quad \frac{2}{\pi}\,\text{p.v}\int_0^{+\infty} \frac{u^2}{1-u^2}\,\omega\,[\mathbf{D}^S(\omega u)]\, du\,, \\ &= \quad \frac{2}{\pi}\lim_{\eta\to 0}\{\int_0^{1-\eta} + \int_{1+\eta}^{+\infty}\}\,. \end{aligned} \tag{83}$$

- Constructing the random matrix $[\mathbf{K}^S(0)]$ such that

$$[\mathbf{K}^S(0)] = [L_{K^S(0)}]^T\,[\mathbf{G}_{K^S(0)}]\,[L_{K^S(0)}]\,, \tag{84}$$

in which the deterministic matrix $[L_{K^S(0)}]$ is such that

$$[K^S(0)] = [L_{K^S(0)}]^T\,[L_{K^S(0)}]\,. \tag{85}$$

The random matrix $[\mathbf{G}_{K^S(0)}]$ is a $(N_S \times N_S)$ random matrix belonging to ensemble SG_ε^+ defined in Section 9.1.2 whose probability distribution and generator of independent realizations depend only on dimension N_S and on the dispersion parameter $\delta_{K^S(0)}$ that allows the level of uncertainties to be controlled. Note that random matrix $[\mathbf{G}_{K^S(0)}]$ is independent of random matrix $[\mathbf{G}_{D^S}]$.

- Computing the random matrix $[\mathbf{D}^+]$ such that

$$[\mathbf{D}^+] = -[\widehat{\mathbf{N}}^R(0)] = \frac{2}{\pi}\int_0^{+\infty}[\mathbf{D}^S(\omega)]\, d\omega\,. \tag{86}$$

- Defining the random matrix $[\mathbf{K}_0^S]$ such that

$$[\mathbf{K}_0^S] = [\mathbf{K}^S(0)] + [\mathbf{D}^+]\,. \tag{87}$$

- Constructing the random matrix $[\mathbf{K}^S(\omega)]$ such that

$$[\mathbf{K}^S(\omega)] = [\mathbf{K}_0^S] + [\widehat{\mathbf{N}}^R(\omega)]\,. \tag{88}$$

It must be verified that $\omega \mapsto [\mathbf{K}^S(\omega)]$ is effectively positive definite. In [101], the following sufficient condition is proven in order for $[\mathbf{K}^S(\omega)]$ to be a positive-definite random matrix for all $\omega \geq 0$: if for all real vector $\mathbf{y} = (y_1, \ldots, y_{N_S})$, the random function $\omega \mapsto \mathbf{y}^T\,[\mathbf{D}^S(\omega)]\,\mathbf{y}$ is decreasing on $[0\,, +\infty]$, then for all $\omega \geq 0$, $[\mathbf{K}^S(\omega)]$ is a positive-definite random matrix.

(ii) For $\omega < 0$, the family $\{[\mathbf{D}^S(\omega)], [\mathbf{K}^S(\omega)]\}_{\omega<0}$ is deduced from the family $\{[\mathbf{D}^S(\omega)], [\mathbf{K}^S(\omega)]\}_{\omega\geq 0}$ by using Equation (78).

(iii) A numerical procedure for computing the integrals in Cauchy principal value can be found in [103].

9.4. *Stochastic Modeling of Random Matrix* $[\mathbf{M}]$

Since there is no available information concerning the statistical dependency of $[\mathbf{M}]$ with the other random matrices of the problem, then random matrix $[\mathbf{M}]$ is independent of all the other random matrices. The deterministic matrix $[M]$ is positive definite, and can consequently be written as

$$[M] = [L_M]^T\,[L_M]\,, \tag{89}$$

in which $[L_M]$ is an upper triangular real matrix. Using the nonparametric probabilistic approach of uncertainties, the stochastic model of the positive-definite symmetric random matrix $[\mathbf{M}]$ is then defined by

$$[\mathbf{M}] = [L_M]^T [\mathbf{G}_M] [L_M] , \tag{90}$$

where $[\mathbf{G}_M]$ is a $(N \times N)$ random matrix belonging to ensemble $\mathrm{SG}_\varepsilon^+$ defined in Section 9.1.2, and whose probability distribution and generator of independent realizations depend only on dimension N and on the dispersion parameter δ_M.

9.5. Stochastic Modeling of Random Matrix $[\mathbf{K}]$

Since there is no available information concerning the statistical dependency of $[\mathbf{K}]$ with the other random matrices of the problem, then random matrix $[\mathbf{K}]$ is independent of all the other random matrices. For the stochastic modeling of $[\mathbf{K}]$, two cases must be considered.

- *Closed (sealed wall) acoustic cavity*. In such a case, the symmetric positive matrix $[K]$ is of rank $N-1$ and can then be written as

$$[K] = [L_K]^T [L_K] , \tag{91}$$

  in which $[L_K]$ is a rectangular $(N \times N-1)$ real matrix. Using the nonparametric probabilistic approach of uncertainties, the stochastic model of the positive symmetric random matrix $[\mathbf{K}]$ of rank $N-1$ is then defined [75,91] by

$$[\mathbf{K}] = [L_K]^T [\mathbf{G}_K] [L_K] , \tag{92}$$

  where $[\mathbf{G}_K]$ is a $((N-1) \times (N-1))$ random matrix belonging to ensemble $\mathrm{SG}_\varepsilon^+$ defined in Section 9.1.2 whose probability distribution and generator of independent realizations depend only on dimension $N-1$ and on the dispersion parameter δ_K.

- *Almost-closed (non-sealed wall) acoustic cavity or internal liquid with a free surface*. Matrix $[K]$ is positive definite and thus invertible. Consequently, it can be written as

$$[K] = [L_K]^T [L_K] , \tag{93}$$

  in which $[L_K]$ is an upper triangular $(N \times N)$ real matrix. Using the nonparametric probabilistic approach of uncertainties, the stochastic model of this positive symmetric random matrix yields

$$[\mathbf{K}] = [L_K]^T [\mathbf{G}_K] [L_K] , \tag{94}$$

  where $[\mathbf{G}_K]$ is a $(N \times N)$ random matrix belonging to ensemble $\mathrm{SG}_\varepsilon^+$ defined in Section 9.1.2 whose probability distribution and generator of independent realizations depend only on dimension N and on the dispersion parameter δ_K.

9.6. Stochastic Modeling of Random Matrix $[\mathbf{C}]$

Since there is no available information concerning the statistical dependency of $[\mathbf{C}]$ with the other random matrices of the problem, then random matrix $[\mathbf{C}]$ is independent of all the other random matrices. We use the construction proposed in [75,91] in the context of the nonparametric probabilistic approach. Let us assume that $N_S \geq N$ and that the $(N_S \times N)$ real matrix $[C]$ is such that $[C]\,\mathbf{q} = 0$ implies $\mathbf{q} = 0$. If $N \geq N_S$, the following construction must be applied to $[C]^T$ instead of $[C]$. Using the polar decomposition of rectangular matrix $[C]$, one can write

$$[C] = [R]\,[T] , \tag{95}$$

in which the $(N_S \times N)$ real matrix $[R]$ is such that $[R]^T [R] = [I_N]$ and where the symmetric square matrix $[T]$ is a positive-definite symmetric $(N \times N)$ real matrix. Using the Cholesky decomposition, we then have

$$[T] = [L_T]^T [L_T], \tag{96}$$

in which $[L_T]$ is an upper triangular matrix. The $(N_S \times N)$ real random matrix $[\mathbf{C}]$ is then written as

$$[\mathbf{C}] = [R]\,[\mathbf{T}] \quad , \quad [\mathbf{T}] = [L_T]^T [\mathbf{G}_C]\,[L_T], \tag{97}$$

where $[\mathbf{G}_C]$ is a $(N \times N)$ random matrix belonging to ensemble $\mathrm{SG}_\varepsilon^+$ defined in Section 9.1.2 and whose probability distribution and generator of independent realizations depend only on N_S, N, and the dispersion parameter δ_C.

9.7. Hyperparameter of the Stochastic Reduced-Order Model (SROM) and Stochastic Solver

The dispersion parameter δ—also called hyperparameter—of each random matrix $[\mathbf{G}]$ allows its level of dispersion (statistical fluctuations) to be controlled. The hyperparameters (dispersion parameters) of random matrices $[\mathbf{G}_{M^S}]$, $[\mathbf{G}_{D^S}]$, $[\mathbf{G}_{K^S(0)}]$, $[\mathbf{G}_M]$, $[\mathbf{G}_K]$, and $[\mathbf{G}_C]$ are represented by a vector-valued hyperparameter $\boldsymbol{\delta}$ such that

$$\boldsymbol{\delta} = (\delta_{M^S}\,, \delta_{D^S}\,, \delta_{K^S(0)}\,, \delta_M\,, \delta_K\,, \delta_C)\,, \tag{98}$$

which belongs to an admissible set $\mathcal{C}_\delta$ and which allows the level of uncertainties to be controlled for each matrix introduced in the stochastic reduced-order model. Consequently, if no experimental data are available, then $\boldsymbol{\delta}$ has to be used to analyze the robustness of the solution of the vibroacoustic problem with respect to uncertainties by varying $\boldsymbol{\delta}$ in $\mathcal{C}_\delta$.

For a given value of hyperparameter $\boldsymbol{\delta}$ in $\mathcal{C}_\delta$, there are two major classes of methods for solving the stochastic reduced-order model (SROM) defined by Equations (64)–(69). The first one belongs to the category of the spectral stochastic methods (see [76,77,83]). The second one belongs to the class of stochastic sampling techniques for which the Monte Carlo method is the most popular. Such a method is often called non-intrusive because it offers the advantage of only requiring the availability of classical deterministic codes. It should be noted that the Monte Carlo numerical simulation method (e.g., [104,105]) is a very effective and efficient one because it has the four following advantages:

- it is a non-intrusive method,
- it is adapted to massively parallel computation without any software developments,
- it is such that its convergence can be controlled during the computation,
- the speed of convergence is independent of the dimension.

If experimental data are available, there are several possible methodologies (one is the maximum likelihood method [87]) to identify the optimal values of hyperparameter $\boldsymbol{\delta}$ (these aspects are not given here, and we refer the reader to [75,84]). Several works have been published concerning experimental validation of the nonparametric probabilistic approach of both the computational model-parameter uncertainties and the model uncertainties induced by modeling errors (e.g., [96,106–111]).

10. Experimental Validation with a Complex Computational Vibroacoustic Model of an Automobile

We present an experimental validation of the nonparametric probabilistic approach of uncertainties for a complex computational vibroacoustic model of an automobile [75,110].

Description of the vibroacoustic system. The vibroacoustic system is an automobile of a given type with several optional extra, for which a single mean computational model is developed. The experimental variabilities are due to the manufacturing process and to the optional extra. The objective is to predict the booming noise for which the engine rotation is $[1500, 4800]$ rpm (rotations per minute), corresponding to the frequency band [50, 160] Hz, for which the input forces are applied to the engine supports, and for which the output observation is the acoustic pressure at a given point localized in the acoustic cavity.

Nominal computational model and stochastic reduced-order model. The nominal computational model is a finite element model of the structure and of the acoustic cavity shown in Figure 2. The structure is modeled with 978,733 structural DOFs of displacement, and the acoustic cavity is modeled with 8139 acoustical DOFs of pressure. The structural reduced-order basis is such that $N = 1722$ and the acoustical reduced-order basis is such that $N_f = 57$. The hyperparameters of the stochastic reduced-order model are $\delta = (\delta_M, \delta_D, \delta_K)$ for the structure, $\delta_f = \delta_{M_f} = \delta_{D_f} = \delta_{K_f}$ for the acoustic cavity, and δ_C for the vibroacoustic coupling.

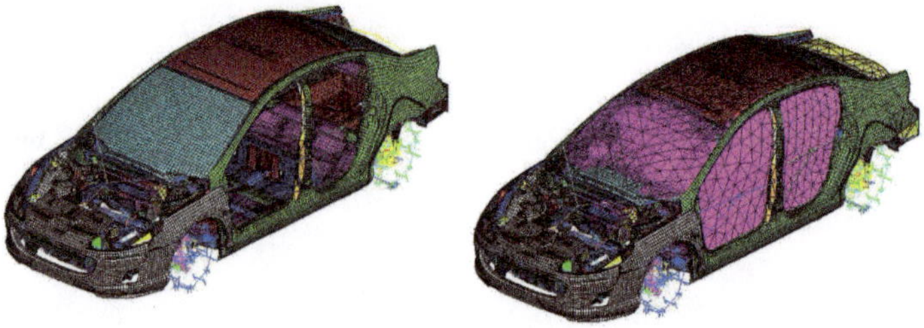

Figure 2. Finite element model of the structure (**left figure**). Finite element mesh of the vibroacoustic computational model (**right figure**). Figure from [110].

Experimental identification of hyperparameter δ_f. The acoustical input is an acoustic source placed inside the acoustic cavity. The acoustical measurements have been performed for $\nu = 30$ cars of the same type with different configurations corresponding to different seat positions, different internal temperatures, and different numbers of passengers. The acoustical pressures have been measured with $\nu_m = 32$ microphones distributed inside the acoustic cavity. For the statistical inverse problem that is required for performing the experimental identification of hyperparameter δ_f, the observation is the real-valued random variable U defined by

$$U = \int_B \mathrm{dB}(\omega)\, d\omega \quad , \quad \mathrm{dB}(\omega) = 10 \log_{10}\left\{ \frac{1}{p_{\mathrm{ref}}^2} \frac{1}{\nu_m} \sum_{j=1}^{\nu_m} |P_{k_j}(\omega)|^2 \right\} , \tag{99}$$

in which $P_{k_1}(\omega), \ldots, P_{k_{\nu_m}}(\omega)$ are the components of $\mathbf{P}(\omega)$, which correspond to the observed DOFs, and which are computed with the SROM defined by Equations (64)–(69). Let $u^{\mathrm{exp},1}, \ldots, u^{\mathrm{exp},\nu}$ be the corresponding measurements for the $\nu = 30$ cars. The identification of hyperparameter δ_f is performed by using the maximum likelihood method [87],

$$\delta_f^{\mathrm{opt}} = \arg\max_{\delta_f} \mathcal{L}(\delta_f) \quad , \quad \mathcal{L}(\delta_f) = \sum_{\ell=1}^{\nu} \log_{10}(p_U(u^{\mathrm{exp},\ell}\,; \delta_f)) . \tag{100}$$

For $\ell = 1, \ldots, \nu$, the value $p_U(u^{\mathrm{exp},\ell}\,; \delta_f)$ of the probability density function (pdf) of the random variable U is estimated with the kernel density estimation method by using the SROM for which the stochastic solver has been chosen as the Monte Carlo method with $\nu_s = 2{,}000$ realizations.

Experimental identification of hyperparameter $\delta = (\delta_{M_s}, \delta_{D_s}, \delta_{K_s})$. The structural inputs are forces applied to the engine supports. The measurements of the accelerations in the structure have been performed for $\nu = 20$ cars of the same type. The random vector-valued observation is $\mathbf{U}(\omega) = (U_1(\omega), \ldots, U_6(\omega))$ such that

$$U_j(\omega) = \log_{10}(\omega^2\, |Y_{k_j}(\omega)|) \quad , \quad j = 1, \ldots, 6 , \tag{101}$$

in which $Y_{k_1}(\omega), \ldots, Y_{k_6}(\omega)$ are the six components of $\mathbf{Y}(\omega)$ which correspond to the observed structural DOFs, and which are computed with the SROM defined by Equations (64)–(69). Let $\mathbf{u}^{\text{exp},1}, \ldots, \mathbf{u}^{\text{exp},\nu}$ be the corresponding measurements for the ν cars. The identification of the hyperparameter $\delta = (\delta_{M_s}, \delta_{D_s}, \delta_{K_s})$ has been performed by using the least-square method. The Monte Carlo method has been used as the stochastic solver with $\nu_s = 1000$ realizations.

Experimental validation. The hyperparameters are fixed to their identified values, $\delta_f = \delta_f^{\text{opt}}$ and $\delta = \delta^{\text{opt}}$, while δ_C is fixed to a given value. The SROM defined by Equations (64)–(69) is solved by using the Monte Carlo method with $\nu_s = 600$ realizations. The prediction—with the identified SROM—of the confidence region of the internal noise at a given point of observation due to the engine excitation is displayed in Figure 3. It can be seen that this prediction is good for representing the great variabilities of the measurements, while the response given by the ROM (the nominal computational model) gives only a rough idea of the real system.

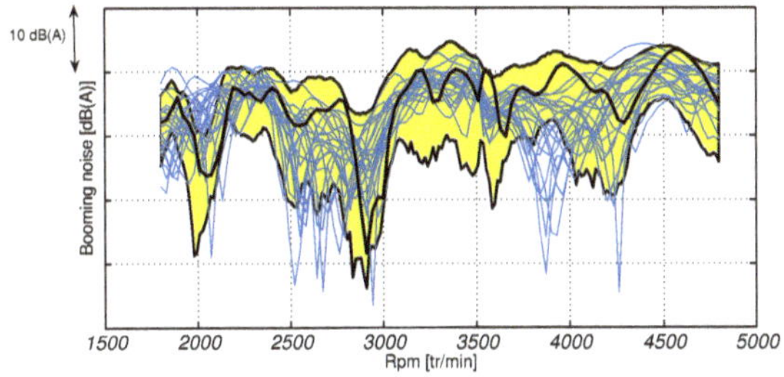

Figure 3. The horizontal axis is the engine rotation expressed in rotations per minute corresponding to the frequency band [50, 160] Hz. The vertical axis is the level of the acoustic pressure in dB(A). Experimental measurements (20 blue thin lines). Prediction with the ROM (black thick line). Prediction with the identified SROM of the confidence region corresponding to the probability level 0.95 (region in yellow color of in gray). Figure from [110].

11. Conclusions

We have presented a computational methodology that is adapted to the vibroacoustics predictions of complex industrial systems in the low- and medium-frequency bands. The vibroacoustic system is made up of a dissipative structure (possibly with a viscoelastic behavior), coupled with a bounded acoustic cavity filled with a dissipative acoustic fluid, and coupled with an acoustic fluid occupying an unbounded external domain. We have presented a reduced-order computational model that is efficient for an acoustic cavity which is filled with a liquid or a gas. For the medium-frequency band, the uncertainty quantification is particularly important. For complex industrial vibroacoustic systems, the main source of uncertainties is due to modeling errors. We have summarized the nonparametric probabilistic approach of model uncertainties, and we have detailed the construction of the random matrices for the case of a viscoelastic structure in order to respect the causality of the physical system. Finally, the application to a real automobile with experimental comparisons was presented.

Author Contributions: The two authors, Roger Ohayon and Christian Soize, have equally contributed on all the aspects of the paper and to write it.

Conflicts of Interest: The authors declare no conflict of interest.

Abbreviations

The following abbreviations and mathematical symbols are used in this manuscript:

BEM	Boundary Element Method
DOF	Degree of Freedom
FEM	Finite Element Method
FRF	Frequency Response Function
FSI	Fluid–Structure Interaction
LF	Low Frequency
MF	Medium Frequency
ROM	Reduced-Order Model
SROM	Stochastic Reduced-Order Model
UQ	Uncertainty Quantification
a_{ijkh}	elastic coefficients of the structure
b_{ijkh}	damping coefficients of the structure
c_0	speed of sound in the internal acoustic fluid
c_E	speed of sound in the external acoustic fluid
$\mathbf{f}$	vector of the generalized forces for the internal acoustic fluid
$\mathbf{f}^S$	vector of the generalized forces for the structure
$\mathbf{g}$	mechanical body force field in the structure
i	imaginary complex number i
k	wave number in the external acoustic fluid
n	number of internal acoustic DOF
n_S	number of structural DOF
n_j	component of vector $\mathbf{n}$
$\mathbf{n}$	outward unit normal to $\partial\Omega$
n_j^S	component of vector $\mathbf{n}^S$
$\mathbf{n}^S$	outward unit normal to $\partial\Omega_S$
p	internal acoustic pressure field
p_E	external acoustic pressure field
$p_E\vert_{\Gamma_E}$	value of the external acoustic pressure field on Γ_E
p_{given}	given external acoustic pressure field
$p_{\text{given}}\vert_{\Gamma_E}$	value of the given external acoustic pressure field on Γ_E
$\mathbf{q}$	vector of the generalized coordinates for the internal acoustic fluid
$\mathbf{q}^S$	vector of the generalized coordinates for the structure
s_{ij}^{damp}	component of the damping stress tensor in the structure
t	time
$\mathbf{u}$	structural displacement field
$\mathbf{v}$	internal acoustic velocity field
x_j	coordinate of point $\mathbf{x}$
$\mathbf{x}$	generic point of $\mathbb{R}^3$
$[A]$	reduced dynamical matrix for the internal acoustic fluid
$[\mathbf{A}]$	random reduced dynamical matrix for the internal acoustic fluid
$[\mathbb{A}]$	dynamical matrix for the internal acoustic fluid
$[A_{\text{BEM}}]$	reduced matrix of the impedance boundary operator for the external acoustic fluid
$[\mathbb{A}_{\text{BEM}}]$	matrix of the impedance boundary operator for the external acoustic fluid
$[A_{\text{FSI}}]$	reduced dynamical matrix for the fluid-structure coupled system
$[\mathbf{A}_{\text{FSI}}]$	random reduced dynamical matrix for the fluid-structure coupled system

$[\mathbb{A}_{\text{FSI}}]$	dynamical matrix for the fluid-structure coupled system
$[A^S]$	reduced dynamical matrix for the structure
$[\mathbf{A}^S]$	random reduced dynamical matrix for the structure
$[\mathbb{A}^S]$	dynamical matrix for the structure
$[A^Z]$	reduced dynamical matrix associated with the wall acoustic impedance
$[\mathbb{A}^Z]$	dynamical matrix associated with the wall acoustic impedance
$[C]$	reduced coupling matrix between the internal acoustic fluid and the structure
$[\mathbf{C}]$	random reduced coupling matrix between the internal acoustic fluid and the structure
$[\mathbb{C}]$	coupling matrix between the internal acoustic fluid and the structure
$[D]$	reduced damping matrix for the internal acoustic fluid
$[\mathbf{D}]$	random reduced damping matrix for the internal acoustic fluid
$[\mathbb{D}]$	damping matrix for the internal acoustic fluid
$[D^S]$	reduced damping matrix for the structure
$[\mathbf{D}^S]$	random reduced damping matrix for the structure
$[\mathbb{D}^S]$	damping matrix for the structure
DOF	degrees of freedom
$\mathbb{F}$	vector of discretized acoustic forces
$\mathbb{F}^S$	vector of discretized structural forces
$G_{ijkh}(0)$	initial elasticity tensor for viscoelastic material
$G_{ijkh}(t)$	relaxation functions for viscoelastic material
$\mathbf{G}$	mechanical surface force field on $\partial\Omega_S$
$[\mathbf{G}]$	random matrix
$[\mathbf{G}_0]$	random matrix
$[K]$	reduced "stiffness" matrix for the internal acoustic fluid
$[\mathbf{K}]$	random reduced "stiffness" matrix for the internal acoustic fluid
$[\mathbb{K}]$	"stiffness" matrix for the internal acoustic fluid
$[K^S]$	reduced stiffness matrix for the structure
$[\mathbf{K}^S]$	random reduced stiffness matrix for the structure
$[\mathbb{K}^S]$	stiffness matrix for the structure
$[M]$	reduced "mass" matrix for the internal acoustic fluid
$[\mathbf{M}]$	random reduced "mass" matrix for the internal acoustic fluid
$[\mathbb{M}]$	"mass" matrix for the internal acoustic fluid
$[M^S]$	reduced mass matrix for the structure
$[\mathbf{M}^S]$	random reduced mass matrix for the structure
$[\mathbb{M}^S]$	mass matrix for the structure
$\mathbb{P}_\alpha$	internal acoustic mode
$[\mathcal{P}]$	matrix of internal acoustic modes
Q	internal acoustic source density
Q_E	external acoustic source density
$\mathbf{Q}$	random vector of the generalized coordinates for the internal acoustic fluid
$\mathbf{Q}^S$	random vector of the generalized coordinates for the structure
$\mathbf{P}$	random vector of internal acoustic pressure DOF
$\mathbb{P}$	vector of internal acoustic pressure DOF
$\mathbf{U}$	random vector of structural displacement DOF
$\mathbb{U}$	vector of structural displacement DOF
$\mathbb{U}_\alpha$	elastic structural mode α
$[\mathcal{U}]$	matrix of elastic structural modes
Z	wall acoustic impedance
$\mathbf{Z}_{\Gamma_E}$	impedance boundary operator for external acoustic fluid
δ	dispersion parameter
ε_{kh}	component of the strain tensor in the structure
ω	circular frequency in rad/s
ρ_0	mass density of the internal acoustic fluid
ρ_E	mass density of the external acoustic fluid
ρ_S	mass density of the structure

σ	stress tensor in the structure
σ_{ij}	component of the stress tensor in the structure
$\sigma_{ij}^{\text{elas}}$	component of the elastic stress tensor in the structure
τ	damping coefficient for the internal acoustic fluid
$\partial\Omega$	boundary of Ω
$\partial\Omega_E$	boundary of Ω_E equal to Γ_E
$\partial\Omega_S$	boundary of Ω_S
Γ	coupling interface between the structure and the internal acoustic fluid
Γ_E	coupling interface between the structure and the external acoustic fluid
Γ_Z	coupling interface between the structure and the internal acoustic fluid with acoustic properties
Ω	internal acoustic fluid domain
Ω_E	external acoustic domain
Ω_S	structural domain

Appendix A. Frequency-Dependent Constitutive Equation for the Dissipative Structure

In this Appendix, we present an appropriate modeling of the frequency-dependent constitutive equation for a dissipative structure in distinguishing the LF range and the MF range [70]. Two cases of frequency-dependent linear constitutive equations are considered in order to describe all the various types of mechanical behaviors encountered in a complex structure. The first one is relevant to the framework of the general linear viscoelasticity theory for describing the constitutive equation of viscoelastic materials, and therefore the frequency-dependent coefficients are constructed in this framework that ensures the causality physical property. This constitutive equation will be referred to as the linear viscoelastic constitutive equation. The second one allows different types of mechanical damping to be modeled using the same type of constitutive equation. The frequency-dependent coefficients will not be constructed in the framework of the linear viscoelasticity theory, but will be constructed in such a way that causality physical property will still be satisfied. This constitutive equation will be referred to as the linear dissipative constitutive equation for modeling damping effects.

Appendix A.1. Linear Viscoelastic Constitutive Equation in the Frequency Domain

The general theory of linear viscoelasticity is used (see [112–114]). With respect to the presentation detailed in [41], we present here a summary of those results with additional developments. In this section, $\mathbf{x}$ is fixed in Ω_S and will be omitted in all the quantities. The Latin indices, such as i, j, k, and h, take the values 1, 2, and 3. The convention for summations over repeated Latin indices is used. The general constitutive equation in the frequency range is written as

$$\sigma_{ij}(\omega) = (a_{ijkh}(\omega) + i\omega\, b_{ijkh}(\omega))\; \varepsilon_{kh}(\omega)\,. \tag{A1}$$

It can be proven that

$$\lim_{|\omega|\to+\infty} a_{ijkh}(\omega) = a_{ijkh}(+\infty)\,, \tag{A2}$$

$$\lim_{|\omega|\to+\infty} \omega\; b_{ijkh}(\omega) = 0\,, \tag{A3}$$

in which $a_{ijkh}(+\infty)$ is called the initial elasticity tensor. It can be deduced that

$$\sigma_{ij}(\infty) = a_{ijkh}(+\infty)\, \varepsilon_{kh}(\infty)\,. \tag{A4}$$

Equation (A4) shows that viscoelastic materials behave elastically at high frequencies with elasticity coefficients defined by the initial elasticity tensor $a_{ijkh}(+\infty)$ that differs from the equilibrium modulus tensor $a_{ijkh}(0)$ that is such that

$$\sigma_{ijkh}(0) = a_{ijkh}(0)\,\varepsilon_{ijkh}(0)\,, \tag{A5}$$

in which $\sigma_{ijkh}(0) = \{\sigma_{ijkh}(\omega)\}_{\omega=0}$ and $\varepsilon_{ijkh}(0) = \{\varepsilon_{ijkh}(\omega)\}_{\omega=0}$. The reader should be aware of the fact that the constitutive equation of an elastic material in a static deformation process is defined by equilibrium modulus tensor $a_{ijkh}(0)$, and not by the initial elasticity tensor $a_{ijkh}(+\infty)$. Referring to [112,115], it has been proven that $a_{ijkh}(0) - a_{ijkh}(+\infty)$ is a negative-definite tensor. The tensors $a_{ijkh}(\omega)$ and $b_{ijkh}(\omega)$ are even functions:

$$a_{ijkh}(-\omega) = a_{ijkh}(\omega) \quad , \quad b_{ijkh}(-\omega) = b_{ijkh}(\omega)\,. \tag{A6}$$

Due to the symmetry properties of tensors $\mathcal{G}_{ijkh}(t)$, it can directly be deduced that tensors $a_{ijkh}(\omega)$ and $b_{ijkh}(\omega)$ must satisfy the symmetry properties

$$a_{ijkh}(\omega) = a_{jikh}(\omega) = a_{ijhk}(\omega) = a_{khij}(\omega)\,, \tag{A7}$$
$$b_{ijkh}(\omega) = b_{jikh}(\omega) = b_{ijhk}(\omega) = b_{khij}(\omega)\,. \tag{A8}$$

In addition, the following positive-definiteness properties can be shown. For all second-order real symmetric tensors X_{ij},

$$a_{ijkh}(\omega)\,X_{kh}\,X_{ij} \geq c_a(\omega)\,X_{ij}\,X_{ij}\,, \tag{A9}$$
$$b_{ijkh}(\omega)\,X_{kh}\,X_{ij} \geq c_b(\omega)\,X_{ij}\,X_{ij}\,, \tag{A10}$$

in which the positive constants $c_a(\omega)$ and $c_b(\omega)$ are such that $c_a(\omega) \geq c_0 > 0$ and $c_b(\omega) \geq c_0 > 0$, where c_0 is a positive real constant independent of ω.

Appendix A.2. Compatibility Equation between a_{ijkh} and b_{ijkh}

We recall that $a_{ijkh}(0)$ is the equilibrium modulus tensor that is the elastic tensor and which is denoted by a^{elas}_{ijkh},

$$a^{\mathrm{elas}}_{ijkh} = a_{ijkh}(0)\,. \tag{A11}$$

Due to the causality property in the time domain and using the Hilbert transform for causal function [102,116–119]), it can be proven that there is a compatibility equation between a_{ijkh} and b_{ijkh}, also called the Kramers and Kronig relation (see [120,121]), which is written

$$a_{ijkh}(\omega) = a^{\mathrm{elas}}_{ijkh} + \frac{\omega}{\pi}\ \mathrm{p.v}\int_{-\infty}^{+\infty} \frac{b_{ijkh}(\omega')}{\omega - \omega'}\,d\omega'\,, \tag{A12}$$

in which p.v denotes the Cauchy principal value. If $y \mapsto h(y)$ is a locally integrable function on the real line except in a singular point $y = 0$, then the p.v is defined as

$$\mathrm{p.v}\int_{-\infty}^{+\infty} h(y)\,dy = \lim_{\ell\to+\infty,\eta\to 0^+} \{\int_{-\ell}^{-\eta} h(y)\,dy + \int_{\eta}^{\ell} h(y)\,dy\}\,. \tag{A13}$$

Appendix A.3. Construction of the Linear Viscoelastic Constitutive Equation in the Frequency Domain

Two cases are considered.

- *(i) Particular case.* A family of linear viscoelastic constitutive equations can be constructed in the time domain using linear differential equations in $\sigma(t)$ and $\varepsilon(t)$. The associated frequency-dependent coefficients $a_{ijkh}(\omega)$ and $b_{ijkh}(\omega)$ automatically verified Equation (A12). In this framework, some examples for $a_{ijkh}(\omega)$ and $b_{ijkh}(\omega)$ can be found in the literature (e.g., [41,98,112,113,122–127]).

- *(ii) General case.* In the general case for which $a_{ijkh}(\omega)$ and $b_{ijkh}(\omega)$ are not derived from such an algebraic representation but correspond to a general integral operator in the time domain (e.g., constructed using experimental curves), a rigorous method of construction is proposed below to satisfy the causality principle.

For the general case, it is assumed that a part Ω_{visco} of the structure Ω_S is made of material modeled in the framework of the linear viscoelasticity theory (see after), while the complementary part Ω_{damp} will be modeled with a linear dissipative constitutive equation for modeling damping effects (detailed in Section A.4). We then have $\Omega_S = \Omega_{\text{visco}} \cup \Omega_{\text{damp}}$.

For the practical construction of the constitutive equation related to Ω_{visco}, it is assumed that functions $\omega \mapsto b_{ijkh}(\mathbf{x}, \omega)$ for $\omega \geq 0$ and equilibrium modulus tensor $a^{\text{elas}}_{ijkh}(\mathbf{x})$ (which is the symmetric and positive-definite elastic tensor) are given. For real ω and for $\mathbf{x}$ belonging to Ω_{visco}, functions $\omega \mapsto a_{ijkh}(\mathbf{x}, \omega)$ can then be constructed.

- The given functions $\omega \mapsto b_{ijkh}(\mathbf{x}, \omega)$ cannot be arbitrarily chosen, but must satisfy some hypotheses to ensure the coherence of the viscoelastic model:

  (1) For all fixed $\mathbf{x}$ and ω, the tensor $\{b_{ijkh}(\mathbf{x}, \omega)\}_{ijkh}$ must be symmetric and positive definite.

  (2) For $\omega \to +\infty$, Equation (A3) must hold, which means that functions $b_{ijkh}(\mathbf{x}, \omega)$ decrease at infinity at least in $\omega^{-\alpha}$ with $\alpha > 1$.

  (3) Functions $\omega \mapsto b_{ijkh}(\mathbf{x}, \omega)$ that satisfy (1) and (2) are then extended to $\omega < 0$ using the even property defined by Equation (A6).

- For all fixed $\mathbf{x}$ and ω, the tensor $\{a_{ijkh}(\mathbf{x}, \omega)\}_{ijkh}$ must be symmetric and positive definite. For all $\omega \geq 0$, functions $\omega \mapsto a_{ijkh}(\mathbf{x}, \omega)$ are then constructed using the following equation (see Equation (A12)),

$$a_{ijkh}(\mathbf{x}, \omega) = a^{\text{elas}}_{ijkh}(\mathbf{x}) + \frac{\omega}{\pi}\, \text{p.v} \int_{-\infty}^{+\infty} \frac{b_{ijkh}(\mathbf{x}, \omega')}{\omega - \omega'}\, d\omega' \,, \tag{A14}$$

  and are extended to $\omega < 0$ using the even property.

- As seen above, for all fixed $\mathbf{x}$ and ω, symmetric tensor $\{a_{ijkh}(\mathbf{x}, \omega)\}_{ijkh}$ must be positive definite. This property must then be checked at the end of the construction, and if it is not satisfied, functions $\omega \mapsto b_{ijkh}(\mathbf{x}, \omega)$ must be modified. In [101], it has been shown that the following sufficient condition allows this property to be satisfied: if functions $\omega \mapsto b_{ijkh}(\mathbf{x}, \omega)$ are decreasing functions for $\omega \geq 0$, then the property is verified.

Appendix A.4. Linear Dissipative Constitutive Equation for Modeling Damping Effects

This section deals with the linear dissipative constitutive equation for modeling damping effects in the part Ω_{damp} of the structure Ω_S. Several models of dissipative constitutive equation (with frequency-dependent coefficients) corresponding to an elastic material are considered for which the mechanical damping is arbitrarily introduced in order to represent damping effects. The first model presented is the constitutive equation for an elastic material with a linear viscous damping term. The second one is a constitutive equation for an elastic material with a parameterized family of damping models depending on frequency. The construction proposed for these two models is such that the causality principle will be verified, and consequently, the fourth-order tensor $a_{ijkh}(\omega)$ will depend on ω although the elastic tensor a^{elas}_{ijkh} of the elastic material is independent of ω.

(i) Constitutive equation for an elastic material with a linear viscous damping term. In this case, the constitutive equation is given by Equation (A1), in which

$$a_{ijkh}(\mathbf{x}, \omega) = a^{\text{elas}}_{ijkh}(\mathbf{x}), b_{ijkh}(\mathbf{x}, \omega) = b_{ijkh}(\mathbf{x}) \,, \tag{A15}$$

in which the tensors $a^{\text{elas}}_{ijkh}(\mathbf{x})$ and $b_{ijkh}(\mathbf{x})$ are symmetric positive definite and independent of ω.

(ii) Constitutive equation for an elastic material with a parameterized family of damping models depending on frequency. The constitutive equation is then defined as an elastic material with a parameterized family of damping models depending on frequency and is written as

$$b_{ijkh}(\mathbf{x},\omega) = \chi(\omega)\, a^{\text{elas}}_{ijkh}(\mathbf{x})\,, \tag{A16}$$

in which the tensor $a^{\text{elas}}_{ijkh}(\mathbf{x})$ is symmetric positive definite and where $\chi(\omega)$ is a positive-valued real function in ω which must satisfy the following properties:

(1) For $\omega \to +\infty$, function $\chi(\omega)$ must decrease at infinity at least in $\omega^{-\alpha}$ in which $\alpha > 1$.
(2) Function χ is even, $\chi(-\omega) = \chi(\omega)$.

From Equation (A14), it can be deduced that for all fixed $\mathbf{x}$ and for all $\omega \geq 0$, the symmetric positive definite tensor must be constructed by the following equation:

$$a_{ijkh}(\mathbf{x},\omega) = \left\{ 1 + \frac{\omega}{\pi}\,\text{p.v}\int_{-\infty}^{+\infty} \frac{\chi(\omega')}{\omega-\omega'}\,d\omega' \right\} a^{\text{elas}}_{ijkh}(\mathbf{x})\,, \tag{A17}$$

in order to ensure the causality property for the constitutive equation. For $\omega < 0$, the values of $a_{ijkh}(\mathbf{x},\omega)$ are obtained using the even property in ω of functions $a_{ijkh}(\mathbf{x},\omega)$. Finally, function $\chi(\omega)$ must be such that for all fixed $\mathbf{x}$ and ω, symmetric tensor $\{a_{ijkh}(\mathbf{x},\omega)\}_{ijkh}$ is positive definite. As previously explained, this property will be satisfied if function $\chi(\omega)$ is a decreasing function for $\omega \geq 0$.

Appendix B. Boundary Element Method for the External Acoustic Fluid

The general references related to the boundary element methods and to their discretization can be found in [27–48,98]. The inviscid acoustic fluid occupies the infinite three-dimensional domain Ω_E whose boundary $\partial\Omega_E$ is Γ_E (see Figure A1).

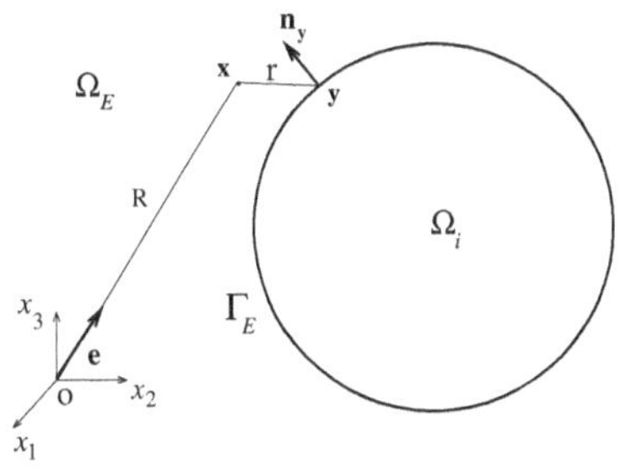

Figure A1. Geometry of the external infinite domain.

This section is devoted to the construction of the frequency-dependent impedance boundary operator $\mathbf{Z}_{\Gamma_E}(\omega)$ such that $p_E|_{\Gamma_E}(\omega) = \mathbf{Z}_{\Gamma_E}(\omega)\,v(\omega)$, which relates the pressure field $p_E|_{\Gamma_E}(\omega)$ exerted by the external fluid on Γ_E to the normal velocity field $v(\omega)$ induced by the deformation of the boundary Γ_E. Furthermore, most of those formulations yield non-symmetric fully populated complex matrices. The computational cost can then be reduced using the fast multipole methods [35,40,44–46]. A major drawback of the classical boundary integral formulations for the exterior Neumann problem related to the Helmholtz equation is related to the uniqueness problem, although the boundary value problem has a unique solution for all real frequencies [97,98]. Precisely, there is not a unique solution of the physical problem for a sequence of real frequencies called spurious or irregular frequencies, also called Jones eigenfrequencies [31,128–131], and various methods are proposed in the literature to overcome this mathematical difficulty arising in the boundary element method [128,132–136]. In this appendix, we present a boundary element method that was initially developed in [137] and detailed in [41]. The formulation of this method is symmetric, and is valid for all real values of the frequency (i.e., without spurious frequencies).

Appendix B.1. Exterior Neumann Boundary Value Problem Related to the Helmholtz Equation

The geometry is defined in Figure A1. The equations of the exterior Neumann problem related to the Helmholtz equation (see Equations (1)–(3)) are rewritten in terms of a velocity potential $\psi(\mathbf{x},\omega)$. Let $\mathbf{v}(\mathbf{x},\omega) = \boldsymbol{\nabla}\psi(\mathbf{x},\omega)$ be the velocity field of the fluid. The acoustic pressure $p(\mathbf{x},\omega)$ is related to $\psi(\mathbf{x},\omega)$ by the following equation:

$$p(\mathbf{x},\omega) = -i\omega\,\rho_{\mathrm{E}}\,\psi(\mathbf{x},\omega) \quad \text{in} \quad \Omega_E\,, \tag{A18}$$

where ρ_{E} is the constant mass density of the external fluid at equilibrium. Let c_{E} be the constant speed of sound in the external fluid at equilibrium, and let $k = \omega/c_{\mathrm{E}}$ be the wave number at frequency ω. The exterior Neumann problem is written as

$$\nabla^2\psi(\mathbf{x},\omega) + k^2\,\psi(\mathbf{x},\omega) = 0 \quad \text{in} \quad \Omega_E\,, \tag{A19}$$

$$\frac{\partial\psi(\mathbf{y},\omega)}{\partial\mathbf{n}_{\mathbf{y}}} = v(\mathbf{y}) \quad \text{on} \quad \Gamma_E\,, \tag{A20}$$

$$|\,\psi\,| = O(\frac{1}{R}) \quad , \quad \left|\frac{\partial\psi}{\partial R} + i\,k\,\psi\right| = O(\frac{1}{R^2})\,, \tag{A21}$$

with $R = \|\mathbf{x}\| \to +\infty$, where $\partial/\partial R$ is the derivative in the radial direction and where $v(\mathbf{y})$ is the given normal velocity field on Γ_E. Equation (A19) is the Helmholtz equation in the external acoustic fluid, Equation (A20) is the Neumann condition on external fluid–structure interface Γ_E, and Equation (A21) corresponds to the outward Sommerfeld radiation condition at infinity. For arbitrary real $\omega \neq 0$, the boundary value problem defined by Equations (A19)–(A21) admits a unique solution denoted ψ^{sol} that depends linearly on the normal velocity v [97,98].

Appendix B.2. Introduction of the Acoustic Impedance Boundary Operator and Radiation Impedance Operator

Let $\psi^{\mathrm{sol}}_{\Gamma_E}$ be the value of ψ^{sol} on Γ_E. For all $\mathbf{x}$ in Ω_E, let $\mathbf{R}(\mathbf{x},\omega/c_{\mathrm{E}})$ be the complex linear operator such that

$$\psi^{\mathrm{sol}}(\mathbf{x},\omega) = \mathbf{R}(\mathbf{x},\omega/c_{\mathrm{E}})\,v\,, \tag{A22}$$

and let $\mathbf{B}_{\Gamma_E}(\omega/c_{\mathrm{E}})$ be the complex linear boundary operator such that

$$\psi^{\mathrm{sol}}_{\Gamma_E} = \mathbf{B}_{\Gamma_E}(\omega/c_{\mathrm{E}})\,v\,. \tag{A23}$$

Using Equation (A18), for all $\mathbf{x}$ in Ω_E, the pressure field $p(\mathbf{x},\omega)$ is written as

$$p(\mathbf{x},\omega) = \mathbf{Z}_{\mathrm{rad}}(\mathbf{x},\omega)\,v\,, \tag{A24}$$

in which $\mathbf{Z}_{\mathrm{rad}}(\mathbf{x},\omega)$ is called the radiation impedance operator that can then be written as

$$\mathbf{Z}_{\mathrm{rad}}(\mathbf{x},\omega) = -i\,\omega\,\rho_{\mathrm{E}}\,\mathbf{R}(\mathbf{x},\omega/c_{\mathrm{E}})\,. \tag{A25}$$

Similarly, the pressure field $p|_{\Gamma_E}(\omega)$ on Γ_E is written as

$$p|_{\Gamma_E}(\omega) = \mathbf{Z}_{\Gamma_E}(\omega)\,v\,, \tag{A26}$$

in which $\mathbf{Z}_{\Gamma_E}(\omega)$ is called the acoustic impedance boundary operator, which can then be written as

$$\mathbf{Z}_{\Gamma_E}(\omega) = -i\,\omega\,\rho_{\mathrm{E}}\,\mathbf{B}_{\Gamma_E}(\omega/c_{\mathrm{E}})\,. \tag{A27}$$

Appendix B.3. Algebraic Properties of the Acoustic Impedance Boundary Operator

Let ${}^t\mathbf{B}_{\Gamma_E}(\omega/c_{\mathrm{E}})$ be the transpose of complex operator $\mathbf{B}_{\Gamma_E}(\omega/c_{\mathrm{E}})$. We have the following symmetry property:

$${}^t\mathbf{B}_{\Gamma_E}(\omega/c_{\mathrm{E}}) = \mathbf{B}_{\Gamma_E}(\omega/c_{\mathrm{E}})\,, \tag{A28}$$

and from Equation (A27), it can be deduced that

$${}^t\mathbf{Z}_{\Gamma_E}(\omega) = \mathbf{Z}_{\Gamma_E}(\omega)\,. \tag{A29}$$

It should be noted that these complex operators are symmetric but not Hermitian. Operator $i\omega\mathbf{Z}_{\Gamma_E}(\omega)$ can be written as

$$i\omega\mathbf{Z}_{\Gamma_E}(\omega) = -\omega^2\,\mathbf{M}_{\Gamma_E}(\omega/c_{\mathrm{E}}) + i\omega\mathbf{D}_{\Gamma_E}(\omega/c_{\mathrm{E}}), \tag{A30}$$

in which $\mathbf{M}_{\Gamma_E}(\omega/c_{\mathrm{E}})$ and $\mathbf{D}_{\Gamma_E}(\omega/c_{\mathrm{E}})$ are two real linear operators such that

$$\omega\,\mathbf{M}_{\Gamma_E}(\omega/c_{\mathrm{E}}) = \Im m\,\mathbf{Z}_{\Gamma_E}(\omega)\,, \tag{A31}$$

$$\mathbf{D}_{\Gamma_E}(\omega/c_{\mathrm{E}}) = \Re e\,\mathbf{Z}_{\Gamma_E}(\omega)\,. \tag{A32}$$

The real part $\mathbf{D}_{\Gamma_E}(\omega/c_{\mathrm{E}})$ of the acoustic impedance boundary operator is symmetric and positive, due to the Sommerfeld radiation condition at infinity.

Appendix B.4. Construction of the Acoustic Impedance Boundary Operator for All Real Values of the Frequency

This construction is based on the use of two boundary integral equations on Γ_E. The first one is based on the use of a single- and double-layer potentials on Γ_E. The second is obtained by the taking the normal derivative on Γ_E of the first one. We then obtained the following system relating $\psi^{\mathrm{sol}}_{\Gamma_E}$ to v, which then allows $\mathbf{B}_{\Gamma_E}(\omega/c_{\mathrm{E}})$ to be defined using Equation (A23),

$$\begin{bmatrix} 0 \\ \mathbf{I}\,\psi^{\mathrm{sol}}_{\Gamma_E} \end{bmatrix} = \begin{bmatrix} -\mathbf{S}_{\mathrm{T}}(\omega/c_{\mathrm{E}}) & \frac{1}{2}\,{}^t\mathbf{I} - {}^t\mathbf{S}_{\mathrm{D}}(\omega/c_{\mathrm{E}}) \\ \frac{1}{2}\,\mathbf{I} - \mathbf{S}_{\mathrm{D}}(\omega/c_{\mathrm{E}}) & \mathbf{S}_{\mathrm{S}}(\omega/c_{\mathrm{E}}) \end{bmatrix} \begin{bmatrix} \psi_{\Gamma_E} \\ v \end{bmatrix}. \tag{A33}$$

The linear boundary integral operators $\mathbf{S}_{\mathrm{S}}(\omega/c_{\mathrm{E}})$, $\mathbf{S}_{\mathrm{D}}(\omega/c_{\mathrm{E}})$, and $\mathbf{S}_{\mathrm{T}}(\omega/c_{\mathrm{E}})$ are defined by

$$< \mathbf{S}_{\mathrm{S}}(\omega/c_{\mathrm{E}})\,v\,,\delta v > = \int_{\Gamma_E}\int_{\Gamma_E} G(\mathbf{x}-\mathbf{y})\,v(\mathbf{y})\,\delta v(\mathbf{x})\,ds_{\mathbf{y}}\,ds_{\mathbf{x}}\,, \tag{A34}$$

$$< \mathbf{S}_{\mathrm{D}}(\omega/c_{\mathrm{E}})\,\psi_{\Gamma_E}\,,\delta v > = \int_{\Gamma_E}\int_{\Gamma_E} \frac{\partial G(\mathbf{x}-\mathbf{y})}{\partial \mathbf{n}_{\mathbf{y}}}\,\psi_{\Gamma_E}(\mathbf{y})\,\delta v(\mathbf{x})\,ds_{\mathbf{y}}\,ds_{\mathbf{x}}\,, \tag{A35}$$

$$\begin{aligned} < \mathbf{S}_{\mathrm{T}}(\omega/c_{\mathrm{E}})\psi_{\Gamma_E},\delta\psi_{\Gamma_E} > = & -k^2\int_{\Gamma_E}\int_{\Gamma_E} G(\mathbf{x}-\mathbf{y})\,\mathbf{n}_{\mathbf{x}}\cdot\mathbf{n}_{\mathbf{y}}\,\psi_{\Gamma_E}(\mathbf{y})\,\delta\psi_{\Gamma_E}(\mathbf{x})\,ds_{\mathbf{y}}\,ds_{\mathbf{x}} \\ & + \int_{\Gamma_E}\int_{\Gamma_E} G(\mathbf{x}-\mathbf{y})\,\{\mathbf{n}_{\mathbf{y}}\times\nabla_{\mathbf{y}}\psi_{\Gamma_E}(\mathbf{y})\}\cdot\{\mathbf{n}_{\mathbf{x}}\times\nabla_{\mathbf{x}}\delta\psi_{\Gamma_E}(\mathbf{x})\}\,ds_{\mathbf{y}}\,ds_{\mathbf{x}}\,, \end{aligned} \tag{A36}$$

where $G(\mathbf{x}-\mathbf{y})$ is the Green function, which is written as

$$G(\mathbf{x}-\mathbf{y}) = g(\|\mathbf{x}-\mathbf{y}\|) = -(4\pi)^{-1} e^{-ikr}/r\,, \tag{A37}$$

in which $r = \|\mathbf{x}-\mathbf{y}\|$. In Equations (A34)–(A36), the brackets correspond to bilinear forms which allow the operators to be defined, and the functions δv and $\delta\psi_{\Gamma_E}$ are associated with functions v and ψ_{Γ_E}. Considering Equation (A33), let $\mathbf{H}(\omega/c_{\mathrm{E}})$ be the operator defined by

$$\mathbf{H}(\omega/c_{\mathrm{E}}) = \begin{bmatrix} -\mathbf{S}_{\mathrm{T}}(\omega/c_{\mathrm{E}}) & \frac{1}{2}\,{}^t\mathbf{I} - {}^t\mathbf{S}_{\mathrm{D}}(\omega/c_{\mathrm{E}}) \\ \frac{1}{2}\,\mathbf{I} - \mathbf{S}_{\mathrm{D}}(\omega/c_{\mathrm{E}}) & \mathbf{S}_{\mathrm{S}}(\omega/c_{\mathrm{E}}) \end{bmatrix}. \tag{A38}$$

Operator $\mathbf{H}(\omega/c_{\mathrm{E}})$ has the symmetric property, ${}^t\mathbf{H}(\omega/c_{\mathrm{E}}) = \mathbf{H}(\omega/c_{\mathrm{E}})$. In Equation (A33), the first equation can be rewritten as $\mathbf{S}_{\mathrm{T}}(\omega/c_{\mathrm{E}})\,\psi_{\Gamma_E} = (\frac{1}{2}\,{}^t\mathbf{I} - {}^t\mathbf{S}_{\mathrm{D}}(\omega/c_{\mathrm{E}}))\,v$. This classical boundary equation that allows the velocity potential to be calculated for a given normal velocity has a unique solution for all real ω, which does not belong to the set of frequencies for which $\mathbf{S}_{\mathrm{T}}(\omega/c_{\mathrm{E}})$ has a null space that is not reduced to $\{0\}$. This set of frequencies is called the set of spurious frequencies. Consequently, for a frequency that is a spurious frequency, ψ_{Γ_E} is the sum of solution $\psi^{\mathrm{sol}}_{\Gamma_E}$ with an arbitrary element of the null space of $\mathbf{S}_{\mathrm{T}}(\omega/c_{\mathrm{E}})$. The method then consists of using the second equation that is written as $\psi^{\mathrm{sol}}_{\Gamma_E} = (\frac{1}{2}\,\mathbf{I} - \mathbf{S}_{\mathrm{D}}(\omega/c_{\mathrm{E}}))\,\psi_{\Gamma_E} + \mathbf{S}_{\mathrm{S}}(\omega/c_{\mathrm{E}})\,v$, and which yields solution $\psi^{\mathrm{sol}}_{\Gamma_E}$ for all real ω. Concerning the practical construction of $\psi^{\mathrm{sol}}_{\Gamma_E}$, for all real values of ω, using Equation (A33), a particular elimination procedure is described in Section B.6.

Appendix B.5. Construction of the Radiation Impedance Operator

The solution $\{\psi^{\mathrm{sol}}(\mathbf{x},\omega), \mathbf{x}\in\Omega_E\}$ of Equations (A19)–(A21) can be calculated using the following integral equation:

$$\psi^{\mathrm{sol}}(\mathbf{x},\omega) = \int_{\Gamma_E} \{G(\mathbf{x}-\mathbf{y})\,v(\mathbf{y}) - \psi^{\mathrm{sol}}_{\Gamma_E}(\mathbf{y},\omega)\,\frac{\partial G(\mathbf{x}-\mathbf{y})}{\partial \mathbf{n}_{\mathbf{y}}}\}\,ds_{\mathbf{y}}\,. \tag{A39}$$

For all $\mathbf{x}$ fixed in Ω_E, the linear integral operators $\mathbf{R}_{\mathrm{S}}(\mathbf{x},\omega/c_{\mathrm{E}})$ and $\mathbf{R}_{\mathrm{D}}(\mathbf{x},\omega/c_{\mathrm{E}})$ are defined by

$$\mathbf{R}_{\mathrm{S}}(\mathbf{x},\omega/c_{\mathrm{E}})\,v = \int_{\Gamma_E} G(\mathbf{x}-\mathbf{y})\,v(\mathbf{y})\,ds_{\mathbf{y}}\,, \tag{A40}$$

$$\mathbf{R}_{\mathrm{D}}(\mathbf{x},\omega/c_{\mathrm{E}})\,\psi_{\Gamma_E} = \int_{\Gamma_E} \psi_{\Gamma_E}(\mathbf{y})\,\frac{\partial G(\mathbf{x}-\mathbf{y})}{\partial \mathbf{n}_{\mathbf{y}}}\,ds_{\mathbf{y}}\,. \tag{A41}$$

Using Equation (A23), Equation (A39) can be rewritten as

$$\psi^{\mathrm{sol}}(\mathbf{x},\omega) = \{\mathbf{R}_{\mathrm{S}}(\mathbf{x},\omega/c_{\mathrm{E}}) - \mathbf{R}_{\mathrm{D}}(\mathbf{x},\omega/c_{\mathrm{E}})\,\mathbf{B}_{\Gamma_E}(\omega/c_{\mathrm{E}})\}\,v\,. \tag{A42}$$

From Equation (A22), it can be deduced that, for all $\mathbf{x}$ in Ω_E,

$$\mathbf{R}(\mathbf{x},\omega/c_{\mathrm{E}}) = \mathbf{R}_{\mathrm{S}}(\mathbf{x},\omega/c_{\mathrm{E}}) - \mathbf{R}_{\mathrm{D}}(\mathbf{x},\omega/c_{\mathrm{E}})\,\mathbf{B}_{\Gamma_E}(\omega/c_{\mathrm{E}})\,, \tag{A43}$$

and the radiation impedance operator $\mathbf{Z}_{\mathrm{rad}}(\mathbf{x},\omega)$ is derived from Equations (A25) and (A43),

$$\mathbf{Z}_{\mathrm{rad}}(\mathbf{x},\omega) = -i\,\omega\,\rho_{\mathrm{E}}\,\{\mathbf{R}_{\mathrm{S}}(\mathbf{x},\omega/c_{\mathrm{E}}) - \mathbf{R}_{\mathrm{D}}(\mathbf{x},\omega/c_{\mathrm{E}})\,\mathbf{B}_{\Gamma_E}(\omega/c_{\mathrm{E}})\}\,. \tag{A44}$$

Appendix B.6. Symmetric Boundary Element Method without Spurious Frequencies

The finite element method is used for discretizing the boundary integral operators $\mathbf{S}_{\mathrm{S}}(\omega/c_{\mathrm{E}})$, $\mathbf{S}_{\mathrm{D}}(\omega/c_{\mathrm{E}})$, and $\mathbf{S}_{\mathrm{T}}(\omega/c_{\mathrm{E}})$ (called a boundary element method). Let us consider a finite element mesh of boundary Γ_E. Let $\mathbf{V} = (V_1,\ldots,V_{n_E})$ and $\mathbf{\Psi}_{\Gamma_E} = (\Psi_{\Gamma_E,1},\ldots,\Psi_{\Gamma_E,n_E})$ be the complex vectors of the n_E

degrees-of-freedom consisting of the values of v and ψ_{Γ_E} at the nodes of the mesh. Let $[S_S(\omega/c_E)]$, $[S_D(\omega/c_E)]$, and $[S_T(\omega/c_E)]$ be the full complex matrices corresponding to the discretization of the operators defined in Equations (A34)–(A36). The complex matrices $[S_S(\omega/c_E)]$ and $[S_T(\omega/c_E)]$ are symmetric. The finite element discretization of Equation (A33) yields

$$\begin{bmatrix} \mathbf{0} \\ [\,E\,]\,\mathbf{\Psi}^{\text{sol}}_{\Gamma_E} \end{bmatrix} = [H(\omega/c_E)] \begin{bmatrix} \mathbf{\Psi}_{\Gamma_E} \\ \mathbf{V} \end{bmatrix}, \tag{A45}$$

in which the symmetric complex matrix $[H(\omega/c_E)]$ is written as

$$\begin{bmatrix} -[S_T(\omega/c_E)] & \frac{1}{2}\,[\,E\,]^T - [S_D(\omega/c_E)]^T \\ \frac{1}{2}\,[\,E\,] - [S_D(\omega/c_E)] & [S_S(\omega/c_E)] \end{bmatrix}. \tag{A46}$$

In Equation (A45), $\mathbf{\Psi}^{\text{sol}}_{\Gamma_E}$ is the complex vector of the nodal unknowns corresponding to the finite element discretization of $\psi^{\text{sol}}_{\Gamma_E}$. The matrix $[\,E\,]$ is the non-diagonal $(n_E \times n_E)$ real matrix corresponding to the discretization of identity operator $\mathbf{I}$. The elimination of $\mathbf{\Psi}_{\Gamma_E}$ in Equation (A45) yields a linear equation between $[\,E\,]\,\mathbf{\Psi}^{\text{sol}}_{\Gamma_E}$ and $\mathbf{V}$, which defines the symmetric $(n_E \times n_E)$ complex matrix $[B_{\Gamma_E}(\omega/c_E)]$, which corresponds to the finite element discretization of boundary integral operator $\mathbf{B}_{\Gamma_E}(\omega/c_E)$,

$$[\,E\,]\,\mathbf{\Psi}^{\text{sol}}_{\Gamma_E} = [B_{\Gamma_E}(\omega/c_E)]\,\mathbf{V}\,. \tag{A47}$$

Vector $\mathbf{\Psi}_{\Gamma_E}$ is eliminated using a Gauss elimination with a partial row pivoting algorithm [138]. If ω does not belong to the set of the spurious frequencies, then $[S_T(\omega/c_E)]$ is invertible and the elimination in Equation (A45) is performed up to row number n_E. If ω coincides with a spurious frequency ω_α (i.e., $\omega = \omega_\alpha$), then $[S_T(\omega_\alpha/c_E)]$ is not invertible, and its null space is a real subspace of $\mathbb{C}^{n_E}$ of dimension $n_\alpha < n_E$. In this case, the elimination in Equation (A45) is performed up to row number $n_E - n_\alpha$. In practice, n_α is unknown. During the Gauss elimination with a partial row pivoting algorithm, the elimination process is stopped when a "zero" pivot is encountered. It should be noted that when the elimination is stopped, the equations corresponding to row numbers $n_E - n_\alpha + 1, \ldots, n_E$ are automatically satisfied. From Equation (A27), we deduce that the $(n_E \times n_E)$ complex symmetric matrix $[Z_{\Gamma_E}(\omega)]$ of operator $\mathbf{Z}_{\Gamma_E}(\omega)$ is such that

$$[Z_{\Gamma_E}(\omega)] = -i\,\omega\,\rho_E\,[B_{\Gamma_E}(\omega/c_E)]\,. \tag{A48}$$

The finite element discretization of the acoustic radiation impedance operator $\mathbf{Z}_{\text{rad}}(\mathbf{x}, \omega)$, defined by Equation (A45), is written as

$$[Z_{\text{rad}}(\mathbf{x}, \omega)] = -i\,\omega\,\rho_E\,\{[R_S(\mathbf{x}, \omega/c_E)] - [R_D(\mathbf{x}, \omega/c_E)]\,[B_{\Gamma_E}(\omega/c_E)]\}\,. \tag{A49}$$

Appendix B.7. Acoustic Response to Prescribed Wall Displacement Field and Acoustic Source Density

We consider the acoustic response of the infinite external acoustic fluid submitted to an external acoustic excitation induced by an acoustic source $Q_E(\mathbf{x}, \omega)$ and to a normal velocity field on Γ_E, which is written as $v = i\omega\,\mathbf{u}(\omega) \cdot \mathbf{n}^S$ (see Section 2 and Figure 1).

(i) Pressure in Ω_E. For all $\mathbf{x}$ in Ω_E, the total pressure $p_E(\mathbf{x}, \omega)$ is written as

$$p_E(\mathbf{x}, \omega) = p_{\text{rad}}(\mathbf{x}, \omega) + p_{\text{given}}(\mathbf{x}, \omega)\,. \tag{A50}$$

The field $p_{\text{rad}}(\mathbf{x},\omega)$ that is radiated by the boundary Γ_E submitted to the given velocity field v is written (see Equation (A24)) as

$$p_{\text{rad}}(\mathbf{x},\omega) = i\omega\,\mathbf{Z}_{\text{rad}}(\mathbf{x},\omega)\{\mathbf{u}(\omega)\cdot\mathbf{n}^S\}\,. \tag{A51}$$

The given pressure $p_{\text{given}}(\mathbf{x},\omega)$ is such that

$$p_{\text{given}}(\mathbf{x},\omega) = p_{\text{inc,Q}}(\mathbf{x},\omega) - \mathbf{Z}_{\text{rad}}(\mathbf{x},\omega)\{\frac{\partial\psi_{\text{inc,Q}}}{\partial\mathbf{n}^S}\}\,, \tag{A52}$$

in which $p_{\text{inc,Q}}(\mathbf{x},\omega)$ is the pressure in the free space induced by the acoustic source Q_E, which is written as

$$p_{\text{inc,Q}}(\mathbf{x},\omega) = -i\omega\int_{K_Q} G(\mathbf{x}-\mathbf{x}')\,Q(\mathbf{x}',\omega)\,d\mathbf{x}'\,, \tag{A53}$$

where G is defined by Equation (A37), and where $\partial\psi_{\text{inc,Q}}/\partial\mathbf{n}^S$ is deduced from Equations (A18) and (A53). In the right-hand side of Equation (A52), the second term corresponds to the scattering of the incident wave induced by the external acoustic source by the boundary Γ_E that is considered as rigid and fixed.

(ii) Pressure on Γ_E. From Equation (A50), it can be deduced that the total pressure on Γ_E is written as

$$p_E|_{\Gamma_E}(\omega) = p_{\text{rad}}|_{\Gamma_E}(\omega) + p_{\text{given}}|_{\Gamma_E}(\omega)\,, \tag{A54}$$

in which $p_{\text{rad}}|_{\Gamma_E}(\omega)$ is such that

$$p_{\text{rad}}|_{\Gamma_E}(\omega) = i\omega\,\mathbf{Z}_{\Gamma_E}(\omega)\{\mathbf{u}(\omega)\cdot\mathbf{n}^S\}\,, \tag{A55}$$

and where the pressure field $p_{\text{given}}|_{\Gamma_E}(\omega)$ on Γ_E is written

$$p_{\text{given}}|_{\Gamma_E}(\omega) = p_{\text{inc,Q}}|_{\Gamma_E}(\omega) - \mathbf{Z}_{\Gamma_E}(\omega)\{\frac{\partial\psi_{\text{inc,Q}}}{\partial\mathbf{n}^S}\}\,. \tag{A56}$$

Equation (A54) can then be rewritten as

$$p_E|_{\Gamma_E}(\omega) = p_{\text{given}}|_{\Gamma_E}(\omega) + i\omega\,\mathbf{Z}_{\Gamma_E}(\omega)\{\mathbf{u}(\omega)\cdot\mathbf{n}^S\}. \tag{A57}$$

Appendix B.8. Asymptotic Formula for the Radiated Pressure Far Field

At point $\mathbf{x}$ in the external domain Ω_E, the radiated pressure $p(\mathbf{x},\omega)$ is given (see Equation (A24)) by $p(\mathbf{x},\omega) = \mathbf{Z}_{\text{rad}}(\mathbf{x},\omega)\,v$. For an observation point $\mathbf{x}$ that is far from the structure, the Green function is rapidly oscillating that induces numerical difficulties for performing the integration over Γ_E. To circumvent this difficulty, asymptotic formulas can be used. Introducing the following parameterization of vector $\mathbf{x}$,

$$\mathbf{x} = R\,\mathbf{e} \quad \text{with} \quad R = \|\mathbf{x}\|\,, \tag{A58}$$

the asymptotic formulas are written as

$$\lim_{R\to+\infty} \mathbf{R}_{\text{S}}(R\mathbf{e}\,;\omega/c_{\text{E}}) = \mathbf{R}_{\text{S}}^{\infty}(R,\mathbf{e}\,;\omega/c_{\text{E}})\,, \tag{A59}$$

$$\lim_{R\to+\infty} \mathbf{R}_{\text{D}}(R\mathbf{e}\,;\omega/c_{\text{E}}) = \mathbf{R}_{\text{D}}^{\infty}(R,\mathbf{e}\,;\omega/c_{\text{E}})\,, \tag{A60}$$

in which the operators $\mathbf{R}_{\text{S}}^{\infty}(R,\mathbf{e}\,;\omega/c_{\text{E}})$ and $\mathbf{R}_{\text{D}}^{\infty}(R,\mathbf{e}\,;\omega/c_{\text{E}})$ are defined by

$$\mathbf{R}_{\text{S}}^{\infty}(R,\mathbf{e}\,;\omega/c_{\text{E}})\,v = \tfrac{1}{R}e^{-i\omega R/c_{\text{E}}}\int_{\Gamma_E} N_{\mathbf{e}}(\mathbf{y})\,v(\mathbf{y})\,ds_{\mathbf{y}}\,, \tag{A61}$$

$$\mathbf{R}_{\mathrm{D}}^{\infty}(R,\mathbf{e}\,;\omega/c_{\mathrm{E}})\,\psi_{\Gamma_{\mathrm{E}}} = \frac{i\omega}{c_{\mathrm{E}}}\,\frac{1}{R}e^{-i\omega R/c_{\mathrm{E}}}\int_{\Gamma_{\mathrm{E}}} \mathbf{e}\cdot\mathbf{n}_{\mathbf{y}}\,N_{\mathbf{e}}(\mathbf{y})\psi_{\Gamma_{\mathrm{E}}}(\mathbf{y})\;ds_{\mathbf{y}}\,, \tag{A62}$$

with

$$N_{\mathbf{e}}(\mathbf{y}) = -\frac{1}{4\pi}\,\exp(i\,\mathbf{e}\cdot\mathbf{y}\,\omega/c_{\mathrm{E}})\,. \tag{A63}$$

From Equation (A43), the asymptotic formula for the radiation impedance operator is then written as

$$\lim_{R\to+\infty} \mathbf{Z}_{\mathrm{rad}}(R\mathbf{e},\omega) = -i\omega\rho_{\mathrm{E}}\{\mathbf{R}_{\mathrm{S}}^{\infty}(R,\mathbf{e}\,;\omega/c_{\mathrm{E}}) - \mathbf{R}_{\mathrm{D}}^{\infty}(R,\mathbf{e}\,;\omega/c_{\mathrm{E}})\,\mathbf{B}_{\Gamma_{E}}(\omega\,c_{\mathrm{E}})\}\,.$$

References

1. Morse, P.M.; Ingard, K.U. *Theoretical Acoustics*; McGraw-Hill: New York, NY, USA, 1968.
2. Lighthill, J. *Waves in Fluids*; Cambridge University Press: Boston, MA, USA, 1978.
3. Cremer, L.; Heckl, M.; Ungar, E.E. *Structure-Born Sound*; Springer: Berlin, Germany, 1988.
4. Pierce, A.D. *Acoustics: An Introduction to Its Physical Principles and Applications*; Originally Published in 1981, McGraw-Hill, New York, NY, USA; Acoustical Society of America Publications on Acoustics: Woodbury, NY, USA, 1989.
5. Crighton, D.G.; Dowling, A.P.; Ffowcs-Williams, J.E.; Heckl, M.; Leppington, F.G. *Modern Methods in Analytical Acoustics*; Springer: Berlin, Germany, 1992.
6. Landau, L.; Lifchitz, E. *Fluid Mechanics*; Pergamon Press: Oxford, UK, 1992.
7. Junger, M.C.; Feit, D. *Sound, Structures and Their Interaction*; Originally Published in 1972, MIT Press, Cambridge, UK; Acoustical Society of America Publications on Acoustics: Woodbury, NY, USA, 1993.
8. Blackstock, D.T. *Fundamentals in Physical Acoustics*; John Wiley & Sons: New York, NY, USA, 2000.
9. Bruneau, M. *Fundamentals of Acoustics*; ISTE USA: Newport Beach, CA, USA, 2006.
10. Fahy, F.J.; Gardonio, P. *Sound and Structural Vibration, Second Edition: Radiation, Transmission and Response*; Academic Press: Oxford, UK, 2007.
11. Howe, M.S. *Acoustics of Fluid-Structure Interactions*; Cambridge Monographs on Mechanics, Cambridge University Press: Cambridge, UK, 2008.
12. Bathe, K.J.; Wilson, E.L. *Numerical Methods in Finite Element Analysis*; Prentice-Hall: New York, NY, USA, 1976.
13. Hughes, T.J.R. *The Finite Element Method: Linear Static and Dynamic Finite Element Analysis*; Dover Publications: New York, NY, USA, 2000.
14. Fish, J.; Belytshko, T. *A First Course in Finite Elements*; John Wiley and Sons: Chichester, UK, 2007.
15. Zienkiewicz, O.C.; Taylor, R.L.; Fox, D.D. *The Finite Element Method For Solid and Structural Mechanics*, 7th ed.; Elsevier, Butterworth-Heinemann: Amsterdam, The Netherlands, 2014.
16. Hughes, T.J.R.; Cottrell, J.A.; Bazilevs, Y. Isogeometric analysis: CAD, finite elements, NURBS, exact geometry and mesh refinement. *Comput. Methods Appl. Mech. Eng.* **2005**, *194*, 4135–4195.
17. Geers, T.L.; Felippa, C.A. Doubly asymptotic approximations for vibration analysis of submerged structures. *J. Acoust. Soc. Am.* **1983**, *173*, 1152–1159.
18. Harari, I.; Hughes, T.J.R. Finite element methods for the Helmholtz equation in an exterior domain: Model problems. *Comput. Methods Appl. Mech. Eng.* **1991**, *87*, 59–96.
19. Givoli, D. *Numerical Methods for Problems in Infinite Domains*; Elsevier: Amsterdam, The Netherlands; London, UK; New York, NY, USA; Tokyo, Japan, 1992.
20. Harari, I.; Grosh, K.; Hughes, T.J.R.; Malhotra, M.; Pinsky, P.M.; Stewart, J.R.; Thompson, L.L. Recent development in finite element methods for structural acoustics. *Arch. Comput. Methods Eng.* **1996**, *3*, 131–309.
21. Astley, R.J. Infinite elements for wave problems: A review of current formulations and assessment of accuracy. *Int. J. Numer. Methods Eng.* **2000**, *49*, 951–976.
22. Farhat, C.; Tezaur, R.; Toivanen, J. A domain decomposition method for discontinuous Galerkin discretizations of Helmholtz problems with plane waves and Lagrange multipliers. *Int. J. Numer. Methods Eng.* **2009**, *78*, 1513–1531.
23. Oden, J.T.; Prudhomme, S.; Demkowicz, L. A posteriori error estimation for acoustic wave propagation. *Arch. Comput. Methods Eng.* **2005**, *12*, 343–389.

24. Bergen, B.; van Genechten, B.; Vandepitte, D.; Desmet, W. An efficient Trefftz-based method for three-dimensional Helmholtz in unbounded domain. *Comput. Model. Eng. Sci.* **2010**, *61*, 155–175.
25. Ihlenburg, F. *Finite Element Analysis of Acoustic Scattering*; Springer: New York, NY, USA, 2013.
26. Farhat, C.; Harari, I.; Hetmaniuk, U. The discontinuous enrichment method for medium-frequency Helmholtz problems with a spatially variable wavenumber. *Comput. Methods Appl. Mech. Eng.* **2014**, *268*, 126–140.
27. Costabel, M.; Stephan, E. A direct boundary integral equation method for transmission problems. *J. Math. Anal. Appl.* **1985**, *106*, 367–413.
28. Kress, R. *Linear Integral Equations*; Springer: New York, NY, USA, 1989.
29. Brebbia, C.A.; Dominguez, J. *Boundary Elements: An Introductory Course*; McGraw-Hill: New York, NY, USA, 1992.
30. Chen, G.; Zhou, J. *Boundary Element Methods*; Academic Press: New York, NY, USA, 1992.
31. Colton, D.L.; Kress, R. *Integral Equation Methods in Scattering Theory*; Krieger Publishing Company: Malabar, FL, USA, 1992.
32. Hackbusch, W. *Integral Equations, Theory and Numerical Treatment*; Birkhauser Verlag: Basel, Switzerland, 1995.
33. Bonnet, M. *Boundary Integral Equation Methods for Solids and Fluids*; John Wiley: New York, NY, USA, 1999.
34. Gaul, L.; Kögl, M.; Wagner, M. *Boundary Element Methods for Engineers and Scientists*; Springer: Heidelberg, Germany; New York, NY, USA, 2003.
35. Schanz, M.; Steinbach, O.E. *Boundary Element Analysis*; Springer: Berlin/Heidelberg, Germany; New York, NY, USA, 2007.
36. Hsiao, G.C.; Wendland, W.L. *Boundary Integral Equations*; Springer: Berlin/Heidelberg, Germany, 2008.
37. Sauter, S.A.; Schwab, C. *Boundary Elements Methods*; Springer: Berlin/Heidelberg, Germany, 2011.
38. Jones, D.S. Integral equations for the exterior acoustic problem. *Q. J. Mech. Appl. Math.* **1974**, *1*, 129–142.
39. Jones, D.S. *Acoustic and Electromagnetic Waves*; Oxford University Press: New York, NY, USA, 1986.
40. Greengard, L.; Rokhlin, V. A fast algoritm for particle simulations. *J. Comput. Phys.* **1987**, *73*, 325–348.
41. Ohayon, R.; Soize, C. *Structural Acoustics and Vibration*; Academic Press: London, UK, 1998.
42. Von Estorff, O.; Coyette, J.P.; Migeot, J.L. Governing formulations of the BEM in acoustics. In *Boundary Elements in Acoustics—Advances and Applications*; Von Estorff, O., Ed.; WIT Press: Southampton, UK, 2000; pp. 1–44.
43. Nedelec, J.C. *Acoustic and Electromagnetic Equations. Integral Representation for Harmonic Problems*; Springer: New York, NY, USA, 2001.
44. Gumerov, N.A.; Duraiswami, R. *Fast Multipole Methods for the Helmholtz Equation in Three Dimension*; Elsevier Ltd.: Amsterdam, The Netherlands, 2004.
45. Bonnet, M.; Chaillat, S.; Semblat, J.F. Multi-level fast multipole BEM for 3-D elastodynamics. In *Recent Advances in Boundary Element Methods*; Manomis, G.D., Polyzos, D., Eds.; Springer: Berlin, Germany, 2009; pp. 15–27.
46. Brunner, D.; Junge, M.; Gaul, L. A comparison of FE-BE coupling schemes for large scale problems with fluid-structure interaction. *Int. J. Numer. Methods Eng.* **2009**, *77*, 664–688.
47. Lee, M.; Park, Y.S.; Park, Y.; Park, K.C. New approximations of external acoustic-structural interactions: Derivation and evaluation. *Comput. Methods Appl. Mech. Eng.* **2009**, *198*, 1368–1388.
48. Chen, L.; Chen, H.; Zheng, C.; Marburg, S. Structural-acoustic sensitivity analysis of radiated sound power using a finite element/discontinuous fast multipole boundary element scheme. *Int. J. Numer. Methods Fluids* **2016**, *82*, 858–878.
49. Ryckelynck, D. A priori hyperreduction method: An adaptive approach. *J. Comput. Phys.* **2005**, *202*, 346–366.
50. Grepl, M.A.; Maday, Y.; Nguyen, N.C.; Patera, A. Efficient reduced-basis treatment of nonaffine and nonlinear partial differential equations. *ESAIM Math. Model. Numer. Anal.* **2007**, *41*, 575–605.
51. Nguyen, N.; Peraire, J. An efficient reduced-order modeling approach for non-linear parametrized partial differential equations. *Int. J. Numer. Methods Eng.* **2008**, *76*, 27–55.
52. Chaturantabut, S.; Sorensen, D.C. Nonlinear model reduction via discrete empirical interpolation. *SIAM J. Sci. Stat. Comput.* **2010**, *32*, 2737–2764.
53. Degroote, J.; Virendeels, J.; Willcox, K. Interpolation among reduced-order matrices to obtain parameterized models for design, optimization and probabilistic analysis. *Int. J. Numer. Methods Fluids* **2010**, *63*, 207–230.

54. Carlberg, K.; Bou-Mosleh, C.; Farhat, C. Efficient non-linear model reduction via a least-squares Petrov-Galerkin projection and compressive tensor approximations. *Int. J. Numer. Methods Eng.* **2011**, *86*, 155–181.
55. Carlberg, K.; Farhat, C. A low-cost, goal-oriented compact proper orthogonal decomposition basis for model reduction of static systems. *Int. J. Numer. Methods Eng.* **2011**, *86*, 381–402.
56. Amsallem, D.; Zahr, M.J.; Farhat, C. Nonlinear model order reduction based on local reduced-order bases. *Int. J. Numer. Methods Eng.* **2012**, *92*, 891–916.
57. Carlberg, K.; Farhat, C.; Cortial, J.; Amsallem, D. The GNAT method for nonlinear model reduction: Effective implementation and application to computational fluid dynamics and turbulent flows. *J. Comput. Phys.* **2013**, *242*, 623–647.
58. Zahr, M.; Farhat, C. Progressive construction of a parametric reduced-order model for PDE-constrained optimization. *Int. J. Numer. Methods Eng.* **2015**, *102*, 1077–1110.
59. Farhat, C.; Avery, P.; Chapman, T.; Cortial, J. Dimensional reduction of nonlinear finite element dynamic models with finite rotations and energy-based mesh sampling and weighting for computational efficiency. *Int. J. Numer. Methods Eng.* **2014**, *98*, 625–662.
60. Amsallem, D.; Zahr, M.J.; Choi, Y.; Farhat, C. Design optimization using hyper-reduced-order models. *Struct. Multidiscip. Optim.* **2015**, *51*, 919–940.
61. Farhat, C.; Chapman, T.; Avery, P. Structure-preserving, stability, and accuracy properties of the Energy-Conserving Sampling and Weighting (ECSW) method for the hyper reduction of nonlinear finite element dynamic models. *Int. J. Numer. Methods Eng.* **2015**, *102*, 1077–1110.
62. Paul-Dubois-Taine, A.; Amsallem, D. An adaptive and efficient greedy procedure for the optimal training of parametric reduced-order models. *Int. J. Numer. Methods Eng.* **2015**, *102*, 1262–1292.
63. Clough, R.W. *Dynamics of Structures*; McGraw-Hill: New York, NY, USA, 1975.
64. Meirovitch, L. *Computational Methods in Structural Dynamics*; Sijthoff and Noordhoff: Alphen aan den Rijn, The Netherlands, 1980.
65. Argyris, J.; Mlejnek, H.P. *Dynamics of Structures*; Elsevier: Amsterdam, The Netherlands, 1991.
66. Morand, H.P.; Ohayon, R. *Fluid Structure Interaction*; Wiley: Chichester, UK, 1995.
67. Craig, R.R., Jr.; Kurdila, A.J. *Fundamentals of Structural Dynamics*; John Wiley and Sons: Hoboken, NJ, USA, 2006.
68. De Klerk, D.; Rixen, D.J.; Voormeeren, S.N. General framework for dynamic substructuring: History, review and classification of techniques. *AIAA J.* **2008**, *46*, 1169–1181.
69. Preumont, A. *Twelve Lectures on Structural Dynamics*; Springer: Dordrecht, The Netherlands, 2013.
70. Ohayon, R.; Soize, C. *Advanced Computational Vibroacoustics—Reduced-Order Models and Uncertainty Quantification*; Cambridge University Press: New York, NY, USA, 2014.
71. Ohayon, R.; Soize, C. Variational-based reduced-order model in dynamic substructuring of coupled structures through a dissipative physical interface: Recent advances. *Arch. Comput. Methods Eng.* **2014**, *21*, 321–329.
72. Peters, H.; Kessissoglou, N.; Marburg, S. Modal decomposition of exterior acoustic-structure interaction problems with model order reduction. *J. Acoust. Soc. Am.* **2014**, *135*, 2706–2717.
73. Geradin, M.; Rixen, D. *Mechanical Vibrations, Third Edition: Theory and Application to Structural Dynamics*; Wiley: Chichester, UK, 2015.
74. Gruber, F.M.; Rixen, D.J. Evaluation of substructure reduction techniques with fixed and free interfaces. *J. Mech. Eng.* **2016**, *62*, 452–462.
75. Soize, C. *Uncertainty Quantification. An Accelerated Course with Advanced Applications in Computational Engineering (Interdisciplinary Applied Mathematics)*; Springer: New York, NY, USA, 2017.
76. Ghanem, R.; Spanos, P.D. *Stochastic Finite Elements: A Spectral Approach*; Springer: New York, NY, USA, 1991.
77. Ghanem, R.; Spanos, P.D. *Stochastic Finite Elements: A Spectral Approach*; Revised Edition; Dover Publications: New York, NY, USA, 2003.
78. Mace, B.; Worden, W.; Manson, G. Uncertainty in Structural Dynamics. *J. Sound Vib.* **2005**, *288*, 431–790.
79. Schueller, G.I. Uncertainties in Structural Mechanics and Analysis-Computational Methods. *Comput. Struct.* **2005**, *83*, 1031–1150.
80. Schueller, G.I. On the treatment of uncertainties in structural mechanics and analysis. *Comput. Struct.* **2007**, *85*, 235–243.

81. Arnst, M.; Ghanem, R. Probabilistic equivalence and stochastic model reduction in multiscale analysis. *Comput. Methods Appl. Mech. Eng.* **2008**, *197*, 3584–3592.
82. Deodatis, G.; Spanos, P.D. Proceedings of the 5th International Conference on Computational Stochastic Mechanics. *Probab. Eng. Mech.* **2008**, *23*, 103–346.
83. LeMaitre, O.P.; Knio, O.M. *Spectral Methods for Uncertainty Quantification with Applications to Computational Fluid Dynamics*; Springer: Heidelberg, Germany, 2010.
84. Soize, C. *Stochastic Models of Uncertainties in Computational Mechanics*; Lecture Notes in Mechanics Series; Engineering Mechanics Institute (EMI) of the American Society of Civil Engineers (ASCE): Reston, VA, USA, 2012.
85. Beck, J.L.; Katafygiotis, L.S. Updating models and their uncertainties. I: Bayesian statistical framework. *J. Eng. Mech.* **1998**, *124*, 455–461.
86. Beck, J.L.; Au, S.K. Bayesian updating of structural models and reliability using Markov chain Monte Carlo simulation. *J. Eng. Mech. ASCE* **2002**, *128*, 380–391.
87. Spall, J.C. *Introduction to Stochastic Search and Optimization*; John Wiley: Hoboken, NJ, USA, 2003.
88. Kaipio, J.; Somersalo, E. *Statistical and Computational Inverse Problems*; Springer: New York, NY, USA, 2005.
89. Soize, C. A nonparametric model of random uncertainties on reduced matrix model in structural dynamics. *Probab. Eng. Mech.* **2000**, *15*, 277–294.
90. Soize, C. Maximum entropy approach for modeling random uncertainties in transient elastodynamics. *J. Acoust. Soc. Am.* **2001**, *109*, 1979–1996.
91. Soize, C. Random matrix theory for modeling uncertainties in computational mechanics. *Comput. Methods Appl. Mech. Eng.* **2005**, *194*, 1333–1366.
92. Jaynes, E.T. Information theory and statistical mechanics. *Phys. Rev.* **1957**, *108*, 171–190.
93. Shannon, C.E. A mathematical theory of communication. *Bell Syst. Tech. J.* **1948**, *27*, 379–423, 623–659.
94. Mehta, M.L. *Random Matrices*; Revised and Enlarged Second Edition; Academic Press: New York, NY, USA, 1991.
95. Soize, C.; Farhat, C. A nonparametric probabilistic approach for quantifying uncertainties in low- and high-dimensional nonlinear models. *Int. J. Numer. Methods Eng.* **2017**, *109*, 837–888.
96. Fernandez, C.; Soize, C.; Gagliardini, L. Fuzzy structure theory modeling of sound-insulation layers in complex vibroacoustic uncertain systems—Theory and experimental validation. *J. Acoust. Soc. Am.* **2009**, *125*, 138–153.
97. Sanchez-Hubert, J.; Sanchez-Palencia, E. *Vibration and Coupling of Continuous Systems. Asymptotic Methods*; Springer: Berlin, Germany, 1989.
98. Dautray, R.; Lions, J.L. *Mathematical Analysis and Numerical Methods for Science and Technology*; Springer: Berlin, Germany, 1992.
99. MSC Nastran, T.M. *Dynamic Analysis User's Guide, Chapter 12—Mid-Frequency Acoustics*; MSC Nastran, MSC Software Cooporation: Newport Beach, CA, USA, 2017.
100. Soize, C. Construction of probability distributions in high dimension using the maximum entropy principle. Applications to stochastic processes, random fields and random matrices. *Int. J. Numer. Methods Eng.* **2008**, *76*, 1583–1611.
101. Soize, C.; Poloskov, I.E. Time-domain formulation in computational dynamics for linear viscoelastic media with model uncertainties and stochastic excitation. *Comput. Math. Appl.* **2012**, *64*, 3594–3612.
102. Papoulis, A. *Signal Analysis*; McGraw-Hill: New York, NY, USA, 1977.
103. Capillon, R.; Desceliers, C.; Soize, C. Uncertainty quantification in computational linear structural dynamics for viscoelastic composite structures. *Comput. Methods Appl. Mech. Eng.* **2016**, *305*, 154–172.
104. Fishman, G. *Monte Carlo: Concepts, Algorithms, and Applications*; Springer: New York, NY, USA, 1996.
105. Rubinstein, R.Y.; Kroese, D.P. *Simulation and the Monte Carlo Method*, 2nd ed.; John Wiley & Sons: New York, NY, USA, 2008.
106. Chebli, H.; Soize, C. Experimental validation of a nonparametric probabilistic model of non homogeneous uncertainties for dynamical systems. *J. Acoust. Soc. Am.* **2004**, *115*, 697–705.
107. Chen, C.; Duhamel, D.; Soize, C. Probabilistic approach for model and data uncertainties and its experimental identification in structural dynamics: Case of composite sandwich panels. *J. Sound Vib.* **2006**, *294*, 64–81.
108. Duchereau, J.; Soize, C. Transient dynamics in structures with nonhomogeneous uncertainties induced by complex joints. *Mech. Syst. Signal Process.* **2006**, *20*, 854–867.

109. Soize, C.; Capiez-Lernout, E.; Durand, J.F.; Fernandez, C.; Gagliardini, L. Probabilistic model identification of uncertainties in computational models for dynamical systems and experimental validation. *Comput. Methods Appl. Mech. Eng.* **2008**, *198*, 150–163.
110. Durand, J.F.; Soize, C.; Gagliardini, L. Structural-acoustic modeling of automotive vehicles in presence of uncertainties and experimental identification and validation. *J. Acoust. Soc. Am.* **2008**, *124*, 1513–1525.
111. Capiez-Lernout, E.; Soize, C.; Mignolet, M.P. Post-buckling nonlinear static and dynamical analyses of uncertain cylindrical shells and experimental validation. *Comput. Methods Appl. Mech. Eng.* **2014**, *271*, 210–230.
112. Truesdell, C. *Encyclopedia of Physics, Vol. VIa/3, Mechanics of Solids III*; Springer: Berlin/Heidelberg, Germany; New York, NY, USA, 1973.
113. Bland, D.R. *The Theory of Linear Viscoelasticity*; Pergamon: London, UK, 1960.
114. Fung, Y.C. *Foundations of Solid Mechanics*; Prentice Hall: Englewood Cliffs, NJ, USA, 1968.
115. Coleman, B.D. On the thermodynamics, strain impulses and viscoelasticity. *Arch. Ration. Mech. Anal.* **1964**, *17*, 230–254.
116. Hahn, S.L. *Hilbert Transforms in Signal Processing*; Artech House Signal Processing Library: Boston, MA, USA, 1996.
117. Pandey, J.N. *The Hilbert Transform of Schwartz Distributions and Applications*; John Wiley & Sons: New York, NY, USA, 1996.
118. King, F.W. *Hilbert Transforms*; Vol 1 and Vol 2, Encyclopedia of Mathematics and Its Applications; Cambridge University Press: Cambridge, UK, 2009.
119. Feldman, M. *Hilbert Transform Applications in Mechanical Vibration*; John Wiley & Sons: New York, NY, USA, 2011.
120. Kronig, R.D. On the theory of dispersion of X-rays. *J. Opt. Soc. Am.* **1926**, *12*, 547–557.
121. Kramers, H.A. La diffusion de la lumière par les atomes, Atti del Congresso Internazionale dei Fisica. In Proceedings of the Transactions of Volta Centenary Congress, Como, Italy, 11–20 September 1927; Volume 2, pp. 545–557.
122. Bagley, R.; Torvik, P. Fractional calculus—A different approach to the analysis of viscoelastically damped struture. *AIAA J.* **1983**, *5*, 741–748.
123. Golla, D.F.; Hughes, P.C. Dynamics of viscoelastic structures—A time domain, finite element formulation. *J. Appl. Mech.* **1985**, *52*, 897–906.
124. Lesieutre, G.A.; Mingori, D. Finite element modeling of frequency-dependent material damping using augmenting thermodynamic fields. *J. Guid. Control Dyn.* **1990**, *13*, 1040–1050.
125. Mc Tavish, D.J.; Hughes, P.C. Modeling of linear viscoelastic space structures. *J. Vib. Acoust.* **1993**, *115*, 103–113.
126. Dovstam, K. Augmented Hooke's law in frequency domain. Three dimensional material damping formulation. *Int. J. Solids Struct.* **1995**, *32*, 2835–2852.
127. Lesieutre, G.A. Damping in structural dynamics. In *Encyclopedia of Aerospace Engineering*; Blockley, R., Shyy, W., Eds.; John Wiley: New York, NY, USA, 2010.
128. Burton, A.J.; Miller, G.F. The application of integral equation methods to the numerical solution of some exterior boundary value problems. *Proc. R. Soc. Lond. Ser. A* **1971**, *323*, 201–210.
129. Jones, D.S. Low-frequency scattering by a body in lubricated contact. *Q. J. Mech. Appl. Math.* **1983**, *36*, 111–137.
130. Luke, C.J.; Martin, P.A. Fluid-solid interaction: Acoustic scattering by a smooth elastic obstacle. *SIAM J. Appl. Math.* **1995**, *55*, 904–922.
131. Jentsch, L.; Natroshvili, D. Non-local approach in mathematical problems of fluid-structure interaction. *Math. Method Appl. Sci.* **1999**, *22*, 13–42.
132. Panich, O.I. On the question of solvability of the exterior boundary value problems for the wave equation and Maxwell's equations. *Russ. Math. Surv.* **1965**, *20*, 221–226.
133. Schenck, H.A. Improved integral formulation for acoustic radiation problems. *J. Acoust. Soc. Am.* **1968**, *44*, 41–58.
134. Mathews, I.C. Numerical techniques for three-dimensional steady-state fluid-structure interaction. *J. Acoust. Soc. Am.* **1986**, *79*, 1317–1325.
135. Amini, S.; Harris, P.J. A comparison between various boundary integral formulations of the exterior acoustic problem. *Comput. Methods Appl. Mech. Eng.* **1990**, *84*, 59–75.

136. Amini, S.; Harris, P.J.; Wilton, D.T. *Coupled Boundary and Finite Element Methods for the Solution of the Dynamic Fluid-Structure Interaction Problem*; Lecture Notes in Engineer; Springer: New York, NY, USA, 1992; Volume 77.
137. Angelini, J.J.; Hutin, P.M. Exterior Neumann problem for Helmholtz equation. Problem of irregular frequencies. *La Recherche Aérospatiale* **1983**, 3, 43–52. (In English)
138. Golub, G.H.; van Loan, C.F. *Matrix Computations*; The Johns Hopkins University Press: Baltimore, MD, USA; London, UK, 1989.

© 2017 by the authors. Licensee MDPI, Basel, Switzerland. This article is an open access article distributed under the terms and conditions of the Creative Commons Attribution (CC BY) license (http://creativecommons.org/licenses/by/4.0/).

MDPI
St. Alban-Anlage 66
4052 Basel, Switzerland
Tel. +41 61 683 77 34
Fax +41 61 302 89 18
http://www.mdpi.com

Applied Sciences Editorial Office
E-mail: applsci@mdpi.com
http://www.mdpi.com/journal/applsci

www.ingramcontent.com/pod-product-compliance
Lightning Source LLC
Chambersburg PA
CBHW051857210326
41597CB00033B/5932